Lecture Notes in Computer Scier

Commenced Publication in 1973
Founding and Former Series Editors:
Gerhard Goos, Juris Hartmanis, and Jan van Leeuwen

César Collazos Andréia Liborio
Cristian Rusu (Eds.)

Human Computer Interaction

6th Latin American Conference, CLIHC 2013
Carrillo, Costa Rica, December 2-6, 2013
Proceedings

 Springer

Volume Editors

César Collazos
Universidad del Cauca
Facultad de Ingeniería Electrónica y Telecomunicaciones
Popayán, Colombia
E-mail: ccollazo@unicauca.edu.co

Andréia Liborio
Federal University of Ceará
CEP 63900-000 Quixadá, CE, Brazil
E-mail: andreia.ufc@gmail.com

Cristian Rusu
Pontificia Universidad Católica de Valparaíso
Escuela de Ingeniería Informática
Casilla 4059, Valparaíso, Chile
E-mail: cristian.rusu@ucv.cl

ISSN 0302-9743   e-ISSN 1611-3349
ISBN 978-3-319-03067-8   e-ISBN 978-3-319-03068-5
DOI 10.1007/978-3-319-03068-5
Springer Cham Heidelberg New York Dordrecht London

Library of Congress Control Number: 2013951516

CR Subject Classification (1998): H.5, H.3-4, I.2.10, I.5, C.2, J.1

LNCS Sublibrary: SL 3 – Information Systems and Application, incl. Internet/Web
and HCI

Typesetting: Camera-ready by author, data conversion by Scientific Publishing Services, Chennai, India

Printed on acid-free paper

Springer is part of Springer Science+Business Media (www.springer.com)

Preface

We would like to extend our warmest welcome to all participants of this international joint conference. We celebrate the 6th edition of the bi-annual Latin American Conference on Human-Computer Interaction (CLIHC 2013).

We are proud to present a very rich program, which reflects many of the interests and endeavors of the HCI community in Latin America. Getting people involved, reviewing, discussing, critiquing, and ultimately selecting the contributions that are included in the program has been the work of an international team based in countries including Brazil, Chile, Colombia, France, Mexico, Panama, Peru, Spain, and United States. We received thirty-one full and two short papers, out of which we selected eleven full papers and fourteen short papers. Considering the full papers, the acceptance rate was about 36 percent.

The program of CLIHC 2013 is organized into thematic sessions that broadly reflect the main areas in which our community has been conducting research: cultural issues, assistive technologies, usability, accessibility, multimodal interfaces, design issues, HCI education, and visualization and evaluation techniques, among others. Technical sessions will be framed by the keynote of a distinguished speaker: Jonathan Grudin, who will share his experience and insights on HCI.

We are very grateful for the commitment and support displayed by the Program Committee, additional reviewers and all the volunteers that made CLIHC 2013 possible. We are certain that the conference provided excellent opportunities for sharing current work in Human-Computer Interaction as well as for establishing and strengthening collaboration links among all participants.

Welcome to Carrillo, Costa Rica, and enjoy the conference!

December 2013
César Collazos
Andréia Liborio
Cristian Rusu

Organization

General Chairs

José Bravo	Castilla La Mancha University, Spain
Sergio F. Ochoa	University of Chile, Chile

CLIHC PC Chairs

César Collazos	Universidad del Cauca, Colombia
Andréia Liborio	Federal University of Ceará, Brazil
Cristian Rusu	Pontificia Universidad Catolica de Valparaiso, Chile

Program Committee

Arnulfo Alanis Garza	Instituto Tecnologico de Tijuana, Mexico
Junia Anacleto	Federal University of So Carlos, Brazil
Clodis Boscarioli	University of Sao Paulo, Brazil
Eduardo Calvillo Gamez	City of San Luis Potosí, Mexico
Cesar Collazos	Universidad del Cauca, Colombia
Clarisse De Souza	PUC-Rio, Brazil
Carla Freitas	Federal University of Rio Grande do Sul, Brazil
Elizabeth Furtado	University of Fortaleza, Brazil
Alex Sandro Gomes	Universidade Federal de Pernambuco, Brazil
Victor M. Gonzalez	Instituto Tecnológico Autónomo de México, Mexico
Toni Granollers	University of Lleida, Spain
Francisco Luis Gutiérrez Vela	Universidad de Granada, Spain
Valeria Herskovic	Pontificia Universidad Católica de Chile, Chile
Juan Pablo Hourcade	University of Iowa, USA
Andréia Libório	Federal University of Ceará, Brazil
Cristiano Maciel	Universidade Federal de Mato Grosso, Brazil
José Antonio Macías Iglesias	Universidad Autónoma de Madrid, Spain
Leonel Vinicio Morales Díaz	Universidad Francisco Marroquín, Guatemala
Mario Alberto Moreno Rocha	Universidad Tecnológica de la Mixteca, Mexico
Alberto L. Morán	UABC, Mexico
Jaime Muñoz-Arteaga	Universidad Autónoma de Aguascalientes, Mexico
Manuel Perez-Quinones	Virginia Tech, USA
José Antonio Pow-Sang	Pontificia Universidad Católica del Perú, Perú
Alberto Raposo	Catholic University of Rio de Janeiro, Brazil
Marcela Rodriguez	UABC, Mexico

Silvana Roncagliolo	Pontificia Universidad Catolica de Valparaiso, Chile
Cristian Rusu	Pontificia Universidad Catolica de Valparaiso, Chile
Virginica Rusu	Universidad de Playa Ancha, Chile
J. Alfredo Sánchez	UDLAP, Mexico
Jaime Sánchez	University of Chile, Chile
Monica Tentori	CICESE, Mexico

Table of Contents

Session 1: Adaptive, Adaptable, Intelligence User Interfaces

Session 2: HCI Design - Methods, Tools and Perspectives

Session 3: HCI Evaluation Methods, Tools and Perspectives

Session 4: HCI Impacts on Society

Session 5: HCI Theories and Theoretical Approaches

Session 6: Interaction, Visualization and Information Processing

Consumption of Profile Information from Heterogeneous Sources to Leverage Human-Computer Interaction

María de Lourdes Martínez-Villaseñor[1] and Miguel González-Mendoza[2]

[1] Universidad Panamericana Campus México, Augusto Rodin 498,
Col. Insurgentes-Mixoac, México, D.F., México
[2] Tecnológico de Monterrey, Campus Estado de México
Carretera Lago de Guadalupe Km 2.5, Atizapán de Zaragoza, Edo. de México, México
lmartine@up.edu.mx,mgonza@itesm.mx

Abstract. Ubiquitous computing brings new challenges to system and application designers. It is not enough to deliver information at any time, at any place and in any form; information must be relevant to the user. Ubiquitous user model interoperability allows enrichment of adaptive systems obtaining a better understanding of the user, but conflict resolution is necessary to deliver the best suited values despite the existence of international standards for different concepts. In this paper, we present the algorithm of conflict resolution to consume of profile information from the ubiquitous user model. We illustrate the enrichment of user models with one elemental concept for human-computer interaction: the language concept.

Keywords: User modeling interoperability, ubiquitous user model, human computer interaction, conflict resolution.

1   Introduction

System designers must take into account that it is not enough to deliver information at any time, at any place and in any form; information must be relevant to the task, background and knowledge of the users [1]. A better understanding of the user helps high functionality applications where general assumptions about the users and stereotypes are not enough for the system to interact cooperatively. Each system and application has valuable but partial information about the user that is worth sharing in order to enrich user models. Although ubiquitous user modeling can improve the usability and usefulness of the human-computer interaction, it is important to decrease the effort associated with creating a user model [1]. We argue that integrating profile information of heterogeneous sources, and enabling ubiquitous user modeling interoperability can leverage human-computer interaction, prevent the user from repeated configurations, and decrease the effort to know the user. Making sense of gathered information from heterogeneous sources entails handling syntactic and semantic heterogeneity, and dealing with different possible conflicts as described in [2]. Syntactic and structural standard language and ontology provide a necessary but not sufficient

C. Collazos, A. Liborio, and C. Rusu (Eds.): CLIHC 2013, LNCS 8278, pp. 1–4, 2013.

condition for exchange. Mediation between concepts is also necessary between to build semantic bridges between representations.

In previous work [3], we presented a framework for ubiquitous user interoperability that enables sharing user model information with a mixed approach to bridge the gap between the mentioned approaches as [4] recommend. In this paper, we present the algorithm of conflict resolution to consume profile information from the ubiquitous user model. We provide an example to illustrate how consumption of even one concept, elemental for human-computer interaction entails handling great syntactic and semantic heterogeneity. An algorithm of conflict resolution and selection of best value to deliver is necessary despite the existence of international standards for this concept. The rest of the paper is organized as follows: in section 2 we present algorithm for the consumption of profile information from a ubiquitous user model to leverage human-computer interaction. We describe our demonstration of concept consumption and results in section 3. We conclude an outline our future work in 4.

2   Consumption of Profile Information from an Ubiquitous User Model to Leverage Human-Computer Interaction

The conflict resolution process fetches the best value available for each concept in consumer´s request. As a precondition of consumption algorithms, the Ubiquitous User Model Interoperability Ontology (U2MIO) is required which contains the user model concept scheme, concept schemes and mappings of previously integrated sources and their instances. U2MIO mediator fetches the value candidates for each concept in the consumer request. The interoperability engine performs a best value selection for each concept in the consumer request. If a value is extracted from concept with an *exactMatch* semantic relation, it is at least considered equivalent and the data type and enumeration constraints (if available) have been checked.

Algorithm 1. Conflict resolution and selection of best values to deliver

```
Require: U2MIO ontology, C_s set of concepts in request
Ensure: V_i* (best values for every required concept)
1:   Receive cd_i from consumer request C_s
2:   for all concepts cd_i in C_s
3:      get restriction facet collection F for cd_i
4:      get concept requested V_i with exactMatch in U2MIO
5:      if V_i is empty
6:         get concept requested V_i with closeMatch in U2MIO
7:      end if
8:      if V_i is not empty
9:         if F is not empty
10:            for all v_j in V_i
11:               for all f_k in F
12:                  if v_j satisfies f_k
13:                     increase restriction satisfaction rcc_j
14:                  end if
15:               end for
```

```
16:              end for
17:                 vᵢ*= vⱼ with max(rccⱼ)
18:        else    comment: no facets to check
19:                 vᵢ*= vⱼ with more recent effective date
20:        end if
21:    else    comment: no value available for this concept
22:        vᵢ*=φ
23:    end if
24: end for
25: Deliver Vᵢ*
```

If this conflict resolution algorithm (algorithm 1) is implemented having XML consumer documents that describe an application user model or a web service description, the compatibility can be checked in the following XML constraining facets *length, minLenght, maxLenght, pattern, enumeration, whiteSpace, maxInclusive, maxExclusive, minInclusive, minExclusive, totalDigits,* and *fractionDigits.*

3　Demonstration of Concept Consumption

We exemplify how the process of conflict resolution delivers interchangeable values despite of the great semantic heterogeneity of *language* concepts of the sources. International standards (ISO 639 language codes for example)[5] help in the identification of languages, but they are not universally adopted. We previously integrated the corresponding concept schemes from Facebook LinkedIn and Google+ to the ubiquitous user model. Even if the concept tags are very similar, frequently the content meaning is only significant to the profile provider. The automatic process of concept alignment established semantic mappings between the language concepts of the social network concept schemes and the ubiquitous user model scheme (U2M).

The process of consumption of profile information is used to enrich the basic demographic information of the Microsoft HealthVault using provider's method *putThings.* Although all concepts of this profile are considered, we focus in the language concept to illustrate the conflict resolution. We show the resulting semantic mapping determined by the process of concept alignment in figure 1. Partial concept schemes focusing in language concepts of each source (Facebook, Linkedin and Google+) and the profile consumer (MS HealthVault) are shown. The green arrows represent relations determined as skos:*exactMatch* (concepts are interchangeable) and red arrows correspond to skos:*closeMatch* (concepts are related) relations. When the profile consumer requests for *language* value, its concept scheme is integrated to the user model determining semantic mappings with U2M concept scheme, and then the process of conflict resolution retrieves concept restrictions in order to deliver the best suited value. In this case, the language concept type refers to a vocabulary of ISO-639-1 (1995) and restricts its value to its content. Although this vocabulary refers to a standard, the code used is already superseded with more recent revisions. The value of Google+ for the language satisfies the type and vocabulary restriction and is therefore delivered.

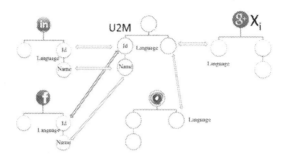

Fig. 1. Language concepts semantic mappings

4   Conclusions and Future Work

We presented the conflict resolution process to fetch the best value selection for requested values from a previously integrated user model. Although standards help when
reusing and sharing profile information, mediation and conflict resolution is needed
when the sources are heterogeneous and autonomous. Conflict resolution process that
exploits the information contained in the source like preferred values, data types and
other restrictions is useful. We illustrate the enrichment of user models with one elemental concept for human-computer interaction: the *language* concept. For future work,
we want to prove that enabling ubiquitous user model interoperability with our framework can decrease the effort of user model design and leverage human-computer
interaction. We are trying to use external vocabularies and ontologies, and considering
situational aspects to make the conflict resolution more accurate, in order to deliver the
best suited values according to the service and user's current situation.

References

1. Fischer, G.: User Modeling in Human-Computer Interaction. User Modeling and User-
   Adapted Interaction 11(1-2), 65–86 (2001)
2. Sosnovsky, S., Brusilovsky, P., Yudelson, M., Mitrovic, A., Mathews, M., Kumar, A.: Semantic Integration of Adaptive Educational Systems. In: Kuflik, T., Berkovsky, S., Car-
   magnola, F., Heckmann, D., Krüger, A. (eds.) Advances in Ubiquitous User Modelling.
   LNCS, vol. 5830, pp. 134–158. Springer, Heidelberg (2009)
3. de Martinez-Villaseñor, M.L.: Design and Implementation of a Framework for Ubiquitous
   User Model Interoperability. Ph.D. Dissertation. Departamento de Posgrados de Ingeniería,
   Instituto Tecnológico y de Estudios Superiores de Monterrey, Campus Estado de México,
   México (2013)
4. Berkovsky, S., Heckmann, D., Kuflik, T.: Addressing Challenges of Ubiquitous User Modeling: Between Mediation and Semantic Integration. In: Kuflik, T., Berkovsky, S., Carmag-
   nola, F., Heckmann, D., Krüger, A. (eds.) Advances in Ubiquitous User Modelling. LNCS,
   vol. 5830, pp. 1–19. Springer, Heidelberg (2009)
5. International Organization for Standardization, Language codes - ISO 639,
   http://www.iso.org/iso/language_codes (accessed May 30, 2013)

Developing Mixed Initiative Educational Web Interfaces for English Education: A Contextual Approach

Marvelia Gizé Jiménez Guzmán[1], Lluvia Morales[2],
Paul Craig[2], and Mario Alberto Moreno Rocha[2]

[1] Universidad del Mar
gize@huatulco.umar.mx
[2] Universidad Tecnológica de la Mixteca
{lluviamorales,p.craig,mmoreno}@mixteco.utm.mx

Abstract. This paper presents early work tackling the problem of developing Mixed Initiative Educational Web Interfaces for English language learning courses that require the adaptation of their contents to different student profiles. The problem is partially solved through a user centered methodology, with our paper focusing on the results of a visual-contextual ethnographic analysis which helped us to identify the user requirements and improve the interactivity, usability and appearance of the interfaces toward developing a true Mixed Initiative system.

Keywords: Mixed Initiative, Adaptive Web Interfaces, User Experience, E-learning.

1   Introduction

The quality of education is a principal driving factor for the development of any country. For this reason it is important that citizens are able to access different educational media, regardless of how they prefer to learn or to which methods they best respond to. In many cases, with this need for flexibility in mind, Communication and Information Technologies (CIT) represent the most feasible option for providing a high-quality low-cost education to the majority of a nation's population [1].

These days, with modern computer technology, it is possible for teachers to create learning activities for their students easily and quickly. These can also be made easily available to students on-line via the internet. This is mainly done through software applications known as Virtual Learning Environments (VLEs) [2] designed to facilitate pedagogical communication between teachers and students, or between students throughout the learning process. Most VLEs are also Learning Content Management Systems (LCMSs) [3] that allow the reuse and sharing of e-learning content held in a central repository. Popular LCMSs include Moodle, Chamilo, WebCT, Sakai and Claroline [4]. These however do not display or allow for the management of learning activities sequences adapted to individual students profiles, and only allow the teacher to develop a generic sequence for a whole group.

C. Collazos, A. Liborio, and C. Rusu (Eds.): CLIHC 2013, LNCS 8278, pp. 5–8, 2013.
© Springer International Publishing Switzerland 2013

A more flexible approach, known as mixed initiative e-learning [5], combines Artificial Intelligence (AI) techniques with LCMSs to adapt learning sequences to individual student needs. But these systems are limited in their capacity to function in an e-learning context, due to the limitations of their interfaces which are not properly adapted for public use. An example of such an interface can be seen in Fig. 1. This is a prime example of an interface developed by software developers *for* software developers with an excessive amount of poorly distributed information making it difficult for users to understand or operate [6]. While these interfaces are often described as being 'mixed initiative', we believe this to be a misnomer since it is difficult for normal users to take the initiative in their part of the interaction due to usability limitations. For this reason we are describing the interface we plan to develop as being a *true* mixed initiative interface. This will be done by employing user centered methodology [7] based on a Visual Ethnographic-Contextual study. As a case study we considered the design of an online English course at our own university.

Fig. 1. A typical Content Planning Interface: Information overload, ill-considered distribution of components and an un-intuitive layout contribute to a poor user experience

2   Case Study

As a case study for this project we plan to implement a mixed initiative on-line course generation/adaptation module for a Virtual Campus e-learning environment that will provide the necessary tools to allow teachers and students to extend both the coverage and scope of teaching/learning while providing a satisfying user experience. As part of an initial requirements analysis for this project we found that undergraduate students in our university did not have a consistent interest in learning English and do not make a proper effort to learn. A sufficient knowledge of English would only become important after a student graduates, or whenever it becomes a mandatory requirement in order to obtain their degree

Part of the objective of the research described in this paper is to tackle this problem through an improved understanding of the requirements and activities of English teachers and students alike. This began with a Visual Ethnographic-Contextual study applying interviews to seven teachers from the Language Center and a Felder and Silverman learning style questionnaires to thirty three PETB (intermediate) level students in order to try and understand their thoughts on English learning and the use of technology.

As a result of this study, our profile of a typical lecturer was foreign lecturer (normally from the United States), aged between twenty-nine and sixty-one years old, with at least a Bachelor's degree and an *English as a second language* teaching certificate.   From the interviews we found that theses teachers considered that the methods and resources for teaching and learning of English in the university were not always relevant. They considered that classes alone were not enough to learn English and they needed to adapt their learning materials to a Mexican context in order to attract students. There was also a feeling that the materials used for learning English were not suitable. The teachers were not able to customize learning materials, and resources did not reflect the students' learning style. They also found that it took too long to mark assignments with the large numbers of students enrolled in the course (30 to 100 students for each teacher). Despite this, all the teachers were found to be motivated to teaching their native language. They would often share advice and discuss their work amongst themselves and they considered technology as a great way to attract the attention and interest of the students to encourage a more active participation in English language learning.

Fig. 2. Interview with an English teacher (left) in her normal working environment.

The students involved in our study were Mexican, aged eighteen to twenty four and studying for a bachelor degree. The results of the questionnaires told us that 65% of the students considered that they *needed* their classes to learn English, 31% considered themselves motivated to learn a new language, and 61% thought that learning English was difficult, 100% of students considered *important* to learn English and 100% used technology on a regular basis, either to listen to music, watch videos or do homework.

3   Proposal

According to the results of our interviews and questionnaires we are beginning to develop Mixed Initiative Learning Management System Interfaces that combine artificial intelligence with teacher input to shape course material to the needs of individual students. This will involve the creation of generic models for an adaptable system to be used in a university environment. This system will be both easy to use and intuitive for lecturers and students alike. It will provide intelligent content and learning activity sequences to lecturers and students and allow lecturers to adapt course content in order to improve the educational experience of the student while elevating some of the logistical problems associated with a larger student to teacher ratio. We aim to assure that these interfaces comply with real world user needs by continuing to enact a usability centered approach to requirements analysis, development and evaluation of software prototypes.

4   Future Work

Due to the need for what we have dubbed as *true* Mixed Initiative e-Learning environments (i.e. those with usable accessible interfaces), we have proposed to generate a number of interfaces prototypes using a user centered methodology. These interfaces will allow teachers to adapt the content and sequence of learning activities for students to improve their learning experience by making course material more relevant while helping resolve workload issues for teachers, by increasing the capacity of students to work independently. Our final prototype will include a full implementation of a mixed initiative system, combining artificial intelligence controlled course sequence planning with human input. This will be evaluated using a full ethnographic study to assess the extent to which a mixed initiative enabled adaptive systems can improve the educational experience of both lecturers and students.

Acknowledgements. We would like to thank the lecturers at the UTM Language Center for their willingness to contribute to this study, the students who took part in the questionnaire and the staff at the UsaLab usability laboratory at the UTM for the use of their facilities and MC Iliana Herrera Arellano for her assistence in English proof reviewing.

References

1. Burbules, N.C., Callister Jr., T.A.: Watch IT: The Risks and Promises of Information Technologies for Education. In: ERIC (2000)
2. Dillenbourg, P., Schneider, D., Synteta, P.: Virtual learning environments. In: Proceedings of the 3rd Hellenic Conference 'Information & Communication Technologies in Education', pp. 3–18 (Year)
3. Robbins, S.R.: The evolution of the learning content management system. Learning Circuits (2002)
4. Kušen, E., Hoic–Bozic, N.: In search of an open–source LMS solution for higher education using a criterion–based approach. International Journal of Learning Technology 7, 115–132 (2012)
5. Garrido, A., Onaindia, E., Morales, L., Castillo, L., Fernández, S., Borrajo, D.: Modeling E-Learning Activities in Automated Planning* (2009)
6. Almenara, J.C.: Bases pedagógicas del e-learning. Revista de Universidad y Sociedad del Conocimiento, RUSC 3, 1 (2006)
7. Nielsen, J.: Usability Engineering. Academic Press Professional, Boston (1993)

Design Choices and Museum Experience: A Design-Based Study of a Mobile Museum App

Olav Røtne and Victor Kaptelinin

Department of Information Science and Media Studies,
University of Bergen, Bergen, Norway
olav.rotne@gmail.com, victor.kaptelinin@infomedia.uib.no

Abstract. The paper reports an experimental study of the effects of visual style, information access selectivity, and content-related challenge on user experience of a mobile museum app prototype. Higher visual richness and added content-related challenge were found to positively affect museum experience, while the effect of information access selectivity was negative.

Keywords: Museum apps, user experience, design dimensions.

1 Introduction

The study reported in this paper explores the effect of a set of design attributes of a mobile museum app on user experience of the app. With the widespread use of personal mobile devices, such as smartphones and tablet computers, museum guides can be implemented as mobile apps [4], which can be downloaded by the visitors. A wide range of museums can be expected to create their own apps. There is a need for HCI research to support this area of practice and help designers find the most efficient ways of using technology to enhance visitors' *experience* [1, 2, 6]. This paper aims to contribute to that effort by presenting empirical evidence about the relationship between certain aspects of a mobile museum app, which are under designer's control (thereafter, "design dimensions") and how the app is experienced by museum visitors. The design dimensions analysed in the study were as follows:

(a) *Visual style* reflected the difference between a refined, "professionally looking" user interface and a less refined, basic graphics design: white space served as the background and hyperlinks (underlined text) were used instead of buttons).

(b) *Information access selectivity* was a degree to which the users could select a concrete information fragment about a museum exhibit, as opposed to viewing all information about an exhibition displayed on one scrollable page.

(c) *Content-related challenge* was achieved by included a multiple-choice "mini quiz", as opposed to providing free access to all information about a museum artefact.

The selection of these dimensions was partly informed by Norman's [3] emotional design model: the selected dimensions roughly correspond to the visceral (Visual style), behavioural (Information access selectivity), and reflective (Content-related challenge) levels of information processing. Experiment 1 was dealing with the first two dimensions, while Experiment 2 focused on the third dimension.

C. Collazos, A. Liborio, and C. Rusu (Eds.): CLIHC 2013, LNCS 8278, pp. 9–13, 2013.
© Springer International Publishing Switzerland 2013

2   Experiment 1

2.1   Method

Eight university students, from 21 to 32 years old, native Norwegian speakers, took part in the study.

The study employed a two-factor within-subject experimental design. Four experimental conditions were produced by combining two levels of the Visual Style variable (Refined vs. Less refined), and two levels of the Information Access Selectivity variable (High selectivity vs. Low selectivity). Examples of user interface designs for the Low selectivity condition are shown in Fig. 1a and Fig. 1b.

The study was conducted at a cultural history division of a major Norwegian museum. The participants were asked to explore four exhibits, Hunting, Harpooning, Agriculture, and Rod Fishing, each comprising a number of stone-age artifacts.

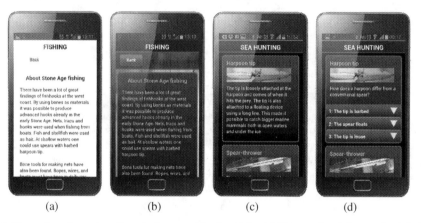

(a)	(b)	(c)	(d)

Fig. 1. Examples of user interface designs used in the study: (a) "Less refined, Low selectivity" condition, (b) "Refined, Low Selectivity" condition, (c) "No challenge" condition, (b) "Mini-quiz" condition. (Text in the images is an English translation of the original Norwegian text)

The participants were tested individually. They were provided with smartphones, which they could use to get access to information about the entire exhibit and individual artifacts. A sequence of four sessions, corresponding to four experimental conditions and involving all four exhibits, was selected for each participant individually according to a Latin Square plan. Each participant also took part in a short interview and completed a survey comprising a number of seven-point Likert scales assessing their experience. The most important scales were "Overall experience" and "Beauty".

2.2   Results

Table 1 shows average "Overall experience" and "Beauty" scores for four experimental conditions of Experiment 1. The results indicate that the "Visual style" dimension had a marked impact on both "Overall experience" and "Beauty" scores in both "Low selectivity" and "High selectivity" conditions. The "Refined" condition was associated with higher scores than the "Less refined" condition for all four interface designs, with the average difference being about 105%. The advantage of the "Refined" condition was supported by the evidence obtained in the interviews.

The results also indicate that, contrary to our expectations, the "Low selectivity" condition was assessed *more* positively: both "Overall experience" and "Beauty" scores for that condition are higher for all four interface designs. The effect is less pronounced than the previous one: the average difference is 22%.

Table 1. Average scores for Experiment 2 conditions, on a scale from "-3" to "+3"

		Information access selectivity			
		Overall experience		**Beauty**	
		Low Selectivity	High Selectivity	Low Selectivity	High Selectivity
Visual Style	Not refined	+1,6	+1,4	+1,3	+0,9
	Refined	+2.6	+2.4	+2.9	+2.4

3   Experiment 2

3.1   Method

Eight university students, from 23 to 29 years old, took part in the study, which employed a one-factor within-subject design. Two experimental conditions corresponded to two levels of the Content-Related Challenge variable: "No challenge" (descriptions of museum artifacts were directly displayed next to pictures of the artefacts) and "Mini-Quiz" (users had to answer multiple-choice questions to get access to a description of an artifact), see Fig 1c and Fig. 1d. The study was conducted in the same setting and using the same museum exhibitions as Experiment 1.

The participants also took part in a short interview and completed a survey comprising a number of seven-point Likert scales. The main scales were "Overall experience", "Learning outcomes", and "Learning motivation".

3.2   Results

Table 2 shows average scores for two experimental conditions of Experiment 2. The results indicate that the "Mini quiz" condition was associated with higher scores than the "No challenge" condition. The difference is manifested in all three scales: "Overall experience", "Learning outcomes", and "Learning motivation", with the average difference being 74%.

Table 2. Average scores for Experiment 2 conditions, on a scale from "-3" to "+3"

		Overall experience	Learning outcomes	Learning motivation
Content-related challenge	No challenge	+1,6	+1,5	+0,9
	Mini quiz	+2,6	+2,2	+1,9

In individual interviews most participants (7 out of 8) stated that the mini quiz made the exhibits more interesting.

4   Discussion of Results and Future Work Directions

The findings from the study allow us to formulate some tentative advices for designers of mobile museum apps. First, the findings suggest that a refined, professionally looking graphical user interface is more important for creating a positive experience than any of the other factors we studied. Therefore, an effort to make the interface look professional can be well justified – especially given that the effort can in principle be rather low (e.g., it can mean choosing a pre-defined template or "skin"). Second, it was found that – as mentioned, contrary to our expectations – providing more advance interactivity, namely, a possibility to selectively choose a specific information fragment describing an exhibit, can be negatively experienced by museum app users. Third, using "mini quizzes" in a multiple choice question format, which required that museum visitors employed their knowledge or inference about museum exhibits, can make a marked positive effect on the experience of museum app users.

The above advices are specifically related to visitor's engagement with particular exhibits, and cannot be directly generalized beyond that scope. For instance, information access selectivity is likely to be associated with positive experience in cases when the user has to choose an information object from a long list of alternatives (e.g., selecting an exhibit to explore rather than a part of exhibit's description). Another limitation of the study is that it only involved a small and homogeneous group of participants, and that it only investigated a subset of design choices that can potentially make an impact on user experience. Further research is needed to understand the role of these factors.

References

1. Gammon, B., Burch, A.: Designing mobile digital experiences. In: Tallon, L., Walker, K. (eds.) Digital Technologies and the Museum Experience. Handheld Guides and other Media. AltaMira Press, Lanham (2008)
2. Kaptelinin, V.: Designing technological support for meaning making in museum learning: An activity-theoretical framework. In: Proceedings of HICSS 44, pp. 1–10. Computer Society Press (2011)

3. Norman, D.: Emotional design: Why we love (or hate) everyday things. Basic Books, NY (2004)
4. Oppermann, R., Specht, M.: Adaptive mobile museum guide for information and learning on demand. In: Proc. HCI International 1999, vol. 2, pp. 642–646 (1999)
5. Røtne. O. Smarttelefon i museet: Et studie av design for brukeropplevelse. Masteroppgave. Universitetet i Bergen (2012) (in Norwegian)
6. Tallon, L., Walker, K. (eds.): Digital Technologies and the Museum Experience. Handheld Guides and other Media. AltaMira Press, Lanham (2008)

Design and Deployment of Everyday UbiComp Solutions at the Hotel: An Empirical Study of Intrinsic Practice Transformation

Rafael Hegre Cabeza and Victor Kaptelinin

Department of Information Science and Media Studies,
University of Bergen, Bergen, Norway
rafael.hegre@gmail.com, victor.kaptelinin@infomedia.uib.no

Abstract. Understanding how people employ digital artifacts in their everyday settings to create more advanced interactive habitats is becoming a key issue in HCI research. This paper aims to contribute to this research by reporting an empirical study of artifact ecologies and their dynamics in day-to-day activities at a hotel. We describe two technological solutions, designed and implemented by people in the settings: (a) converting a paper-based cleaning staff roster into a Google Doc, and (b) switching from a traditional fax machine to email as a technology for handling communication with suppliers. We discuss a range of factors affecting such user-driven innovations, as well as the impact of the technologies on larger-scale interactive habitats.

Keywords: Habitat, end-user development, intrinsic practice transformation, hotel industry, everyday computing, UbiComp.

1 Introduction

With interactive technologies spreading beyond work settings to all areas of our daily life, HCI research has been increasingly focusing on *everyday computing* [1]. This trend is associated with two closely related challenges, that is, understanding and supporting: (a) human interaction with entire ecologies of technology, rather than individual interactive products, and (b) users as designers of their own technology-enabled interactive environments.

The importance of understanding ecologies of technology has been emphasized in a number of recent studies. A variety of concepts, such as place, information ecology, artifact ecology, product ecology, habitat, and so forth (e.g. [2, 6, 9, 10, 11]) have been proposed as conceptual tools for dealing with ecologies of technology rather than particular artifacts. While undeniably useful, such conceptual explorations need to be complemented with more extensive empirical investigations to provide more substantial guidance in analysis, evaluation, and technological support of concrete everyday life contexts and settings.

Expanding the focus of analysis and design beyond the product foregrounds the centrality of users as designers of their own interactive environments [10]. While

C. Collazos, A. Liborio, and C. Rusu (Eds.): CLIHC 2013, LNCS 8278, pp. 14–21, 2013.
© Springer International Publishing Switzerland 2013

individual artifacts comprising an ecology are typically designed by professional designers (including interaction designers), an ecology as a whole is likely to be shaped by influences coming from a variety of stakeholders, with a substantial contribution of people populating the ecology.

The purposeful efforts of technology users, directed at finding and implementing optimal solutions for employing interactive technologies in their own habitats are variously conceptualized in HCI literature as end-user development [5], ephemeral innovations [13], exploring opportunity spaces [8], intrinsic technology-induced practice transformation [10], digital plumbing [14], and users "doing" the UbiComp [12]. A number of important insights have resulted from this research, but further empirical studies are needed to understand and support users acting as designers of everyday UbiComp solutions.

There is a potential obstacle for future progress in this area. One can argue that everyday computing often employs simple low-end technologies, and is therefore less relevant to HCI research, which should be predominantly concerned with innovative interactive artifacts. In our view, this is a misconception. While analysis and design of innovative technologies is, undoubtedly, critically important, it should be complemented with HCI research into the factors and conditions of how people actually make use of the potential of interactive technologies in their everyday lives.

This paper aims to contribute to this research by presenting and analyzing empirical evidence obtained in an ethnographic study conducted at a hotel industry setting. We describe two cases: (a) converting a paper-based cleaning staff roster into a Google Doc, and (b) switching from a traditional fax machine to email as a technology for handling all types of communication with various suppliers. It should be specifically emphasized that the focus of the study was not on using advanced technologies but rather on people appropriate interactive technologies, however simple, in their everyday practices.

The empirical evidence presented in this paper was collected in an ethnographic study conducted by the first author (thereafter referred to as "the researcher"), who was working in the hotel industry setting in question as a manager for extended periods of time. The cases described below represent a subset of the findings generated by the study.

2   The Setting and Actors

The study was conducted in a middle-sized hotel in Norway, predominantly oriented toward business travelers, who typically come during weekdays to spend the night and work the day after. This paper generally discusses two types of actors at the hotel; the cleaning staff manager (hereafter referred to as "the manager") and cleaning staff. The manager has responsibility for the round-the-clock-operations related to maintenance such as cleaning staff, supplier negotiation and cleaning supply inventory. The cleaning staff is responsible for washing designated areas of the hotel, keeping it in a general upkeep so that guests are met with a tidy and sanitary environment.

3   Case A: Roster

3.1   Previous Practice

The roster for the cleaning staff was an A3 paper sheet containing an overview over which shifts each individual will have, over the course of three months (see Figure 1a). It was handled by the manager as the head of the cleaners. Normally she would simply fill in each employee in the days that were needed. If the employees had any requests for how they wanted to work, they gave her written paper notes regarding their wishes, which she then took into consideration.

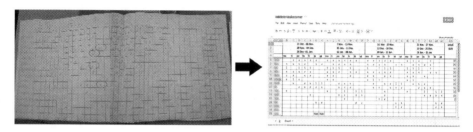

Fig. 1. a) Roster in paper version situated at manager's office, b) Roster in Google Doc format available online

There were three different paper copies of the roster, one at the manager's house, one in her office, and one in the cafeteria for the hotel employees. Usually the roster at the manager's office was changed first and served as a master copy, and the others were then modified accordingly.

Regarding the cleaning staff, the shift information was clearly not ubiquitously available to them, as they did not have the possibility to access the roster from other locations such as their homes. If they for some reason should be uncertain about their shifts, they either had to visit their cafeteria, or call the person in charge of the roster (who either had to be at her office or at home to get access to the roster herself). Sometimes the staff was not able to work the shifts in which they were listed, so that two person's shifts had to be swapped. At one occasion two of the employees needed to change their shifts, and respective changes in the roster involved four people altogether (see Fig. 1a).

To make a shift change the staff called the manager and asked her whether the change was possible. The manager was not able to give a proper answer right away if she was not in the proximity of any of the lists. After actually getting access to the roster, she was able to see what employees were able to work that day and she called them to ask whether they were available. Subsequent to approval, the manager modified rosters in her office and home, while the cleaning staff themselves updated the remaining copy in the cafeteria. Therefore, if one change had to be made, a (marker) pen was used on all three copies at three different venues. A potential problem here was that ambiguity might arise, if something was written differently by different persons. But neither of the actors involved saw any problems with the way this process was handled, and were rather satisfied with it.

3.2   Change of Practice: From Paper to a Google Doc

When discussing with the manager her everyday practices, the researcher pointed to Google Docs as a potential tool for managing the cleaning staff roster. Essentially, Google Docs is a collection of office tools (such as Word and Excel) that are available online via a web browser, and that lets the users have constant access, anytime and anywhere. Therefore, instead of users not knowing whether a roster has been updated at every location, they could now simply access a Google Docs document, and find the relevant roster, even from their laptop or home computer.

A new roster, implemented as a Google Docs spreadsheet, was developed after the researcher showed the manager how the system worked. It should be emphasized that the actual solution was designed and implemented by the manager herself. The new way of implementing the roster was developed in-house by using a relatively familiar tool (spreadsheet), accessed through another known tool, namely the web browser. No radically new technology per se had to be exploited, turning this into a smooth experience and effort.

The manager made the new list almost an exact copy to the one previous used, only the novel roster being digital (see Figure 1a and 1b). After the process of changing the list from a tangible paper format, to a digital presentation in Google Docs, the roster was made available for the cleaning staff by providing a link to the GD document.

After the change of practice, the manager saved time; not only in how the list was written, but also in how it can be accessed. She is now surrounded by the roster, as it is interchangeably available on multiple heterogeneous devices [2]. Her activity space has been changed as the ubiquitous accessibility that the roster now permits allows the modification anytime anywhere as long as she has Internet and some sort of computational device present (this being a stationary computer, laptop, tablet PC, smartphone or the like). She has moved the availability away from the traditional paper presentation, breaking free from a localized tool in her office to rather using portable and mobile technological artifacts. Her transformation has now led her, as well as the cleaning staff, to be more mobile.

4   Case B: From Fax to Email

4.1   Previous Practice

The manager received much of her correspondences from various business partners, such as suppliers, through fax. This involves everything from prices of cleaning supply that they use and what items that are available in stock, to various advertising offers. The manager is currently situated on the second floor, which involves her having to move down to the reception in order to get the printed out faxes, not having a printer present in her office or nearby. Therefore, her activity space [10] was not enhanced properly with technology supporting this particular activity: she did not have the appropriate artifacts (in this case a printer/fax machine) to support the task in a proper manner. But what can be easily seen is that the technological objects available actually could support the activity of receiving information from any supplier, if the processes were only handled somewhat different (see Figure 2 and 3).

Fax-based Correspondence Handling

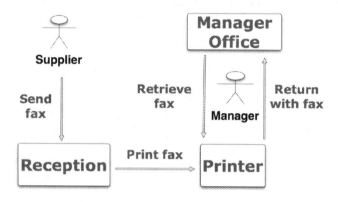

Fig. 2. Previous practice

4.2   Change of Practice: From Fax to Email

As in the case of the cleaning staff roster, described above, the idea for a practice change emerged during a discussion of current practices between the researcher and the manager. The manager wanted an extensive transformation of the practice:

> *"If they [suppliers/other actors] send everything to us via email, we would not need to walk down to the first floor or buy a printer for the second floor, just simply use the computer of our working station" (translated).*

Email-based Correspondence Handling

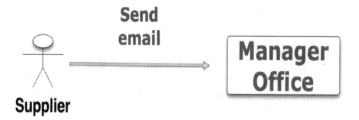

Fig. 3. Current practice

For her the use of fax and printing out of paper became totally incomprehensible, especially since she easily could receive it all on email.

Four months after the conversation surrounding practice, the manager actually transformed the use of fax machine to the use of email, as it had been discussed. This had not been an action realized due to the wishes of hotel management; rather she had executed this based on her own free will and effort. What she had done was to contact all business partners that would usually send her fax correspondences, and asked them to provide the same information through email (this, of course, demanded some effort from the suppliers´ end as well, but was gradually carried out). This has led to a significant change of practice, as she no longer has to spend time and energy on the actual retrieval of paper versions, and can now simply collect them at the click of a button in a digital format. It is not only the faxes from suppliers that now are sent to her via email; also many other related matters that previously were given to her in paper versions, such as messages from colleagues. This is an example of intrinsic practice transformation [8], that is, the manager improving the practice herself by employing the resources that are already in place. She did not need any novel technologies, simply using the email client to exchange digital messages instead of receiving correspondences in paper format.

5   Discussion

There are several aspects of the findings, presented above, which, in our view, deserve special attention.

First, the specific strategies for technology appropriation, as well as necessary technological artifacts (such as the Google Doc roster) were devised and implemented by the practitioner herself. The changes of practice we observed can be partly attributed to the very fact that the study, reported here, took place in the setting. The discussions between the researcher and the practitioner made the latter reflect on problems with existing practices and think about possible solutions (that is, understand the current situation as an opportunity space, see [8]). At the same time, however, neither the researcher nor any other "external" person was involved in the actual design and implementation of a strategy for technology-induced practice transformation. The researcher acted as a facilitator: in informal conversations he helped reflect on existing problems and, in Case A, pointed to a general availability of a certain technology that could be tried out, but he did not act as a co-designer of any of the solutions.

These findings point to the importance of creating conditions for intrinsic practice transformation [10]: in particular, developing effective and efficient strategies for helping practitioners reflect upon their practices, identify potentially useful technologies, and implement their own solutions.

Second, in both cases a solution primarily devised to address a practitioner's own practice had far-reaching effects on the setting (and beyond) and resulted in a reconfiguration of social interactions between the practitioner and other people.

Third, the evidence from our study suggests that practitioners' personalities and attitudes toward technology and change can be a key factor in practice transformation. Some individuals tend to apply solutions to ad hoc problems, that are not necessarily needed or demanded of them, but that still can optimize their practices. These types of supra-situational activities [7] can be significant for a person in transcending her/his requirements. While technology skills are also of importance, they do not have to be "beyond the ordinary".

Fourth, in both cases described in this paper the practitioners employed well-known and rather simple technologies. The novelty of the solutions was relative, with respect to the particular setting, rather than absolute. However, designing and implementing of each of these solutions took a creative effort and was experienced as an individual accomplishment.

Fifths, the changes of practice we observed involved changes in artifact/ product ecologies [3, 6, 9]. Some artifacts (paper-based roster, fax machine) became extinct, while other ones (a Google Doc spreadsheet, email) survived and prospered. In Case A the transformation of artifact ecology was associated with a new artifact entering the scene (the Google Doc-based roster). In Case B the reconfiguration of artifact ecology was achieved by re-positioning a previously existing artifact, that is, by making arrangements that made a relatively peripheral artifact to a central position within the ecology.

Sixth and finally, it would be an oversimplification to assess the changes we observed as entirely positive. While the changes can be considered generally successful, they also had some negative side-effects. For instance, when the online roster was introduced, everyone was forced to use the internet, irrespective of whether or not they wanted to, and switching to email to handle business correspondence, while eliminating the need for the manager to visit another part of the building, also eliminated some possibilities for both formal and informal communication with her colleagues.

References

1. Abowd, G.D., Mynatt, E.D.: Charting past, present, and future research in ubiquitous computing. ACM TOCHI 7(1), 29–58 (2000)
2. Bødker, S., Klokmose, N.: The Human-Artifact Model: An activity theoretical approach to artifact ecologies. Human Computer Interaction 26, 315–371 (2011)
3. Bødker, S., Klokmose, N.: Dynamics in artifact ecologies. In: Proc. NordiCHI 2012: Making Sense Through Design, pp. 448–457 (2012)
4. Carroll, J.M., Rosson, M.B., Farooq, U., Xiao, L.: Beyond being aware. Information and Organization 19(3), 162–185 (2009)
5. Fischer, G., Giaccardi, E., Ye, Y., Sutcliffe, A.G., Mehandjiev, N.: Meta-design: A manifesto for end-user development. CACM 47(9), 33–37 (2004)
6. Forlizzi, J.: The product ecology: Understanding social product use and supporting design culture. International Journal of Design 2(1), 11–20 (2008)
7. Hedestig, U., Kaptelinin, V.: Facilitator's Invisible Expertise and Supra-Situational Activities in a Telelearning Environment. In: Proc. HICSS 2003, pp. 1–10. IEEE Computer Society (2003)

8. Hornecker, E., Halloran, J., Fitzpatrick, G., Weal, M., Millard, D., Michaelides, D., Cruickshank, D., De Roure, D.: UbiComp in opportunity spaces: Challenges for participatory design. In: PDC 2006, pp. 47–56. ACM Press (2006)
9. Jung, H., Stolterman, E., Ryan, W., Thompson, T., Siegel, M.: Toward a framework for ecologies of artifacts: How are digital artifacts interconnected within a personal life? In: Proc. NordiCHI 2008, pp. 201–210. ACM Press (2008)
10. Kaptelinin, V., Bannon, L.: Interaction design beyond the product: Creating technology-enhanced activity spaces. Human-Computer Interaction 27(3), 277–309 (2012)
11. Nardi, B., O'Day, V.: Information ecologies: Using technology with heart. MIT Press, Cambridge (1999)
12. Oulasvirta, A.: When users "do" the UbiComp. Interactions, 7–9 (March/April 2008)
13. Spinuzzi, C.: Tracing genres through organizations: A sociocultural approach to information design. MIT Press, Cambridge (2003)
14. Tolmie, P., Crabtree, A., Rennick Egglestone, S., Humble, J., Greenhalgh, C., Rodden, T.: Digital plumbing: The mundane work of deploying UbiComp in the home. Pers. Ubiquit. Comput. 14(3), 181–196 (2010)

Bringing the Web Closer:
Stereoscopic 3D Web Conversion

Alexey Chistyakov, Diego González-Zúñiga, and Jordi Carrabina

Universitat Autònoma de Barcelona,
Bellatera 08193, Spain
alexey.chistyakov@e-campus.uab.cat,
diekus@acm.org,
jordi.carrabina@uab.cat

Abstract. In this paper we present 3DSjQ, a tool used to implement stereoscopic 3D in web pages. It provides HTML developers the possibility to create static and dynamic content that interacts with depth. We uncover the algorithm used for the tool, describe the method of operation and discuss future work including further development and implementations.

Keywords: stereoscopic, 3D, web, depth, HTML, framework, interface, interaction, javascript.

1 Introduction

Stereoscopy can be used as a narrative technique to try to provoke immersion and arousal in a movie [1]. Despite this, according to the Motion Picture Association of America, only 159 films (2.7% of all released since 2003) have been released in 3D [2]. This aligns with complaints from users regarding available content [3]. But new uses of 3D in education where "marked positive effect of the use of 3D animations on learning" is indicated [4], shed light on the importance of expanding beyond films and allowing the creation of 3D assets and apps. When coupled with forecasts of the rising trend of stereoscopic 3D (S3D) devices [5], a niche for more interactive S3D apps can be noted.

Analysing how much content is on the web, over 672M websites were estimated by Netcraft's June 2013 Web Server survey [6]. If we take the previously exposed situation with S3D movies as a metaphor, we can state that the amount of existing S3D content in relation to the total number of web pages is 0 percent. In order to alleviate this problematic, the idea of converting existing pages into a valid stereoscopic format to be used in stereo displays is of great interest for us. Additionally, future tools or frameworks that allow the introduction of depth in web applications can be valuable when coupled with semantic information. In this article, we present a proposal to do exactly this, convert an existing webpage into a side by side valid stereoscopic format.

C. Collazos, A. Liborio, and C. Rusu (Eds.): CLIHC 2013, LNCS 8278, pp. 22–25, 2013.
© Springer International Publishing Switzerland 2013

2   State of the Technology

Stereoscopy is used mostly in cinematography for entertainment purposes, but three developments are allowing a more interactive approach with 3D: (i) The number of devices equipped with 3D displays is growing [7]. (ii) 3DTVs are becoming more affordable for consumers, and (iii) the development of head mounted displays and the interaction they portray with virtual 3D environments is in expansion.

Related to this, the usage of stereoscopy on the web can be seen in several examples listed here [8]. All of these websites use passive anaglyph images as a background and most of them serve promotional and entertainment purposes, leaving productivity, accessibility and other possible enhancements on a second place, or even unconsidered.

Despite the poor appearance of stereoscopy on the Internet, the World Wide Web Consortium, which is responsible for developing modern web standards, has a proposal called "Extensions for S3D support" [9]. This proposal introduces the extension for CSS (Cascading Style Sheet) properties specific for S3D content.

3   The Tool: 3DSjQ

As a way to approach the problem related to the lack of existing S3D content, we present a tool that adds S3D depth to a web page. In order to achieve this we developed a 2D-to-3D conversion algorithm (referred from now on as 'algorithm'). Following is a description of the process that clones, adapts, mirrors and shifts markup elements on an HTML page in correspondence with the technical guidelines defined by Sky3D [10] using HTML5, CSS3 and jQuery.

3.1   Content Cloning

According to the theory of stereoscopy [1] to achieve the stereoscopic depth illusion, we must create an exact copy of the ``body" part of the existing HTML and paste the cloned markup side by side along the original. Thus the first step of the algorithm is to clone the content. At this stage the script creates two containers for each the original and the clone. Then, it cuts the existing HTML and pastes it into the corresponding containers. After, the output HTML becomes invalid according to the recommendations of W3C which restrict the appearance of elements with the same "id" attribute [11]. In order to revalidate the output markup with the recommendations provided by the W3C, additional processing is executed. The script detects elements with the "id" attributes within the clone container, and adds a prefix which makes their "id" unique.

3.2   Styles Adaptation

Style adaptation is intended to solve two main problems: (i) apply styles for the cloned elements that lost their style rules after their "id" attributes were renamed and (ii) handle the mirroring of interactions defined by dynamic pseudo classes.In order to apply the initial styles to the cloned elements with updated "id" attributes, 3DSjQ uses AJAX (Asynchronous Javascript and XML) to get the stylesheets referenced in the input HTML, parses it with regular expressions, and replaces detected "id" selectors with the prefixed "id" value. It then stores the rule associated with this selector. At the same time the script does the same search for elements with dynamic pseudo classes such as ":hover", ":active", and ":focus". At the end of this part of the conversion process, 3DSjQ collects updated rules and injects them under the "style" tag in the output HTML.

3.3   Interaction Mirroring

During this phase, the main purpose of this script is to apply all the basic interactions from the original part of the content to the clone. It includes the mirroring of the mouse cursor position, basic interactions defined by dynamic pseudo classes and the mirroring of content scrolling.

3.4   zPlane Builder

To create the stereoscopic effect, elements in each container should be shifted to the left or to the right from its original position depending on desirable depth level [1]. Here, 3DSjQ creates an array of data associated with each element specified during the setup. Taking into account all this parameters 3DSjQ shifts and (if visual cues are allowed) scales the elements and its clones from their original positions according to the settings specified along with the initiation function. This way 3D depth illusion is achieved.

4   Conclusion

The tool that converts HTML pages to stereoscopic 3D (S3D) was presented. The processes that compose it have been explained. We developed this tool as a partial solution for the lack of S3D content available on the Internet. We also present this tool as an innovative way to build HTML-based S3D user interfaces for displays and mobile devices that work with this format. The tool is easy to use. It leverages features of open technologies HTML5, CSS3, and jQuery, which are supported and used by the web developer community. The tool was tested to create an HTML page from zero and to convert the Mozilla Foundation homepage and a Google search engine results page. Both cases finished with positive results in all modern browsers and required less than five minutes spent on coding. Nonetheless there are several limitations when interacting with pages that are not valid according to W3C standards.

5   Future Work and Discussion

As of date of writing, the tool is in stable alpha release and can be used as a jQuery plugin on any HTML page. Nevertheless in order to make the tool reliable additional testing, debugging, and code optimization is required.

The tool can also be used to develop static 3D compositions and conduct research on 3D stereoscopic depth perception. 3DSjQ can be used to study user interactions in stereoscopic spatial graphical user interfaces [12], and the influence of stereoscopy on productivity of diverse tasks in education processes [4] and cognitivity.

We also intend to promote 3DSjQ within the developer community. For this, we plan to release a beta version of the script on GitHub and attract more people to participate in further code development in order to improve the quality of the project. At the same time, this will aid the growth of S3D websites on the Internet.

Acknowledgements. Supported by the Spanish Ministry of Finance and Competivity (proj. no. IPT-2012-0630-020000).

References

1. Mendiburu, B.: 3D Movie Making. Stereoscopic Digital Cinema from Script to Screen. Focal Press (2009)
2. Motion Pciture Association of America: Theatrical market statistics 2012 (2013)
3. Nuttall, C.: Lack of content leaves 3d tv sales (2010),
   http://www.ft.com/cms/s/0/13ae797c-093f-11e0-ada6-00144feabdc0.html#axzz2vp4iujxj
4. Bamford, A.: The 3d in education white paper. Technical report, International Research Agency (2011)
5. Hsieh, C.: Market and technology assessment, 3d displays. DisplaySearch (2010)
6. Netcraft: June 2013 web server survey. Technical report, Netcraft,
   http://news.netcraft.com/archives/2013/06/06/june-2013-web-server-survey-3.html
7. Research and Markets: 3d-enabled tv sets on the rise worldwide. Technical report, Research and Markets (2011)
8. Awwwards Team: 10 stereoscopic 3d websites (2011),
   http://www.awwwards.com/10-stereoscopic-3d-websites.html
9. Han, S., Lee, D.Y.: Extensions for stereoscopic 3d support (November 2012),
   http://www.w3.org/2011/webtv/3dweb/3dwebproposal121130.html
10. Sky3D: Bskyb technical guidelines for plano stereoscopic (3d) programme content. Technical report, BSkyB
11. W3C: Html: The markup language (an html language reference) (October 2012),
    http://dev.w3.org/html5/markup/overview.html
12. Bowman, D., Kruijff, E., Laviola, J., Poupyrev, I.: 3D User Interfaces. Theory and Practice. Addison-Wesley Educational Publishers Inc. (2004)

User Experience Degree and Time Restrictions as Financial Constraints in Heuristic Evaluation

Llúcia Masip, Toni Granollers, and Marta Oliva

University of Lleida, Jaume II 69, 25001 Lleida, Spain
{lluciamaar,tonig,oliva}@diei.udl.cat

Abstract. One of the most important concerns of companies is the budget invested in every task of a project. In tech projects, the evaluation of interactive systems is one of the most valuable parts of the development process. And obviously, financial constraints do not avoid this part. In this context, two factors related to heuristic evaluation (one of the most economical methods) can be taken into account: the user experience degree and the time available for the evaluation. A survey with end users (understanding end users those use heuristic evaluation methodology) was carried out to determine the values of both factors in a specific context: website applications.

Keywords: User Experience, Heuristic Evaluation, UX Degree.

1   Introduction

As it is well-known, the budget of a project is the most important factor to be strictly considered in companies. When the estimation of the needed resources is defined, it should be firmly applied in every task of the project. Because the invested project-budget deeply depends on the type of evaluation carried on, this study focuses on the evaluation stage of the development process of an interactive system. Usually, evaluation with end users increases the budget due to the needed time to plan the evaluation and recruit users, contrary to other methodologies without end users, for instance heuristic evaluation (HE) [1] –and we will focus on this methodology. Specifically, this research is focused on the first step of the HE process: the configuration of the evaluation. To be more precise, the important moment of selecting which is the list of heuristics that better fits with the specific interactive system to be evaluated.

2   Financial Constraints in Heuristic Evaluation

There are few projects where scientists carried out strong efforts to classify heuristics in different levels of importance. In [2] the heuristics are classified in three levels: very high, high, medium and low. In another research, the heuristics are classified using a level of importance [3]: the "relative importance" of the guideline and the

C. Collazos, A. Liborio, and C. Rusu (Eds.): CLIHC 2013, LNCS 8278, pp. 26–29, 2013.

"strength of evidence" used in making that judgment. In the following section the concepts of "UX degree" and "time restrictions" are proposed. Then, the survey to reach the UX degree for website application is detailed. Finally, conclusions and future work are given.

2.1   The UX Degree and Time Restrictions

The main goal of the *UX degree (UXD)* is the classification of the whole set of heuristics in more accurate sets to be able to not consider some of them in the case of budget restrictions. It divides into different levels of consideration according to the importance that a specific heuristic has in a specific kind of interactive system.

Therefore, the *UXD represents the level of importance that every heuristic (or full set of heuristics) has in a specific system*. Furthermore, and by imitating accessibility levels [4], three UX degrees are proposed: **U degree**: the heuristics of the U degree are essential to assure that the user who will use the interactive system will get a positive experience. **UU degree**: the heuristics of the UU degree are necessary to assure that the user who will use the interactive system will get a positive experience. **UUU degree**: it is advisable to consider the heuristics of the UUU degree to assure that the user who will use the interactive system will get a positive experience.

Following the accessibility guidelines example, heuristics belonging to U degree are the minimum necessary to consider that who uses the evaluated interactive system will feel a little bit positive experience. But if the interactive system should present a higher level of quality, it would be necessary to consider the three levels.

Furthermore, *"Time is money"*. In business context everyone agrees on the meaning of this saying. Time parameter is not obvious, it depends on the type of interactive system, the experience of the professional in the usage of the inspection methodology and the knowledge that the professionals has about the heuristics (if they are familiar or not with the set of the advised heuristics). But real industrial projects always need to estimate how much time is needed for everything, and user experience evaluations do not escape of this consideration. So, if it is possible to provide project managers with this information, they will be able to deliver a set of heuristics according to the specific budget of this task of the project.

3   Setting Values for the UX Degree and Time Restrictions

The process to determine the values for the UXD and the time restrictions of some interactive system is carried out through a survey. The first goal of the survey is to define a UXD for the heuristics that can be applied in a specific interactive system: website applications. The second goal is to determine the approximate needed time to consider every single heuristic in this specific interactive system.

The set of 267 heuristics was collected in a previous research where all usability definitions from 1986 to now were reviewed [5].

Participants in the survey should be experts in the Human Computer Interaction (HCI) area. So, our option to recruit participants was to send a "Call for participation in a PhD research" to HCI experts, mainly university HCI researchers and UX professionals from different international companies. The call for participation was sent to 79 HCI/UX experts from whom we obtained 63 answers (30 males and 33 females). Users were from 18 to 56 years old and they have at least 2 years of experience (30 users have between 2 and 5 years of experience and 24 users have between 6 and 10 years of experience).

The process that each participant followed to answer the survey was divided in three main steps: (i) Fill in the user profile form. (ii) Answer the first questionnaire selecting the degree that the participant considers more suitable for each heuristic taking into account that the set of heuristics is for a website. (iii) Answer the second questionnaire with questions about the time.

Taking into account the large amount of heuristics, the set of heuristics was divided into 3 groups, considering different facets, to provide the participants with a smaller group of heuristics. Thus, in the first group, the set (Q1) included heuristics from cross-cultural, communicability, findable, accessibility and dependability (80 heuristics). The second (Q2) and the third (Q3) groups included the half amount of usability heuristics (88 and 99 heuristics each one). The usability facet was divided into two groups because the initial set of heuristics was so wide to inspire participants in the answer of the questionnaire. Finally, the 3 different documents were sent to HCI professionals via email. Q1 was sent to 26 people. Q2 was sent to 30 people and Q3 was sent to 23 people. But, unfortunately, not everybody answered the questionnaire. Q1 was answered by 21 participants, Q2 by 18 and Q3 by 24.

4   Results

Due to space limitations the UXD for each heuristic cannot be presented, but can be consulted in the following web address: http://www.grihotools.udl.cat/openheredeux/cake_1_3/uxdegrees/showuxdegree. The specific heuristics that each facet has in each UXD are presented in Table 1:

Table 1. Amount of heuristics for each facet and for each UX degree

UX facet	U	UU	UUU
Cross-cultural	6	5	1
Communicability	19	8	2
Findable	12	1	1
Accessibility	1	1	0
Dependability	15	7	1
Usability	91	75	21

The time restriction factor was asked in the second and last questionnaire. Two questions were asked: "How much time do you think that you need to score one of this heuristics?" and "Is it one minute for evaluating two heuristics enough?" Bearing in mind those 63 experts took part in the research, 42.9% of participants think that less than a minute is enough to score one heuristic. In addition, 66.7% told us that the enough time to score two heuristics is one minute. Therefore, the consideration of one minute to evaluate two heuristics is a good option.

5   Conclusions and Future Work

UXD of heuristics and time restrictions are focused on enhancing the UX evaluation in real cases. It enables a more accurate evaluation scheduling and a much better use of project budget. The definition of a UXD for heuristics that are applicable in web applications is the first step to standardize the UX evaluation. If a consolidation of these results is reached, the certification of the UX will be possible.

UXD and time restrictions are included in Open-HEREDEUX [6] resource. Specifically, in the "Adviser of heuristics" module, where the provided list of heuristics can be refined depending on the UXD and/or time restrictions. It calculates how many heuristics are needed according to the time restriction and selected UXD. But in any case, Adviser recommends at least the whole set of heuristics of U degree because U degree of heuristics is the minimum heuristics that permits to get an interactive system with the minimum level of quality. As a future work, we would like to research about the definition of different weight in heuristics of different UX facets and in heuristics of the same facet. It will provide best budget planning.

Acknowledgments. The work has been supported by Spanish Ministry of Science and Innovation through the InDAGuS-UX (TIN2012-37826-C02-02) and by Universitat de Lleida for pre-doctoral fellowship to Llúcia Masip.

References

1. Nielsen, J., Molich, R.: Heuristic evaluation of user interfaces. In: Proceedings of the SIGCHI Conference on Human Factors in Computing Systems, NY (1990)
2. UserPlus Advisor, http://www.userplus.com
3. Research-Based Web Design & Usability Guidelines. U.S. Dept. of Health and Human Services (2006) ISBN/ASIN: 0160762707
4. W3C. Web Content Accessibility Guidelines 2.0. W3C Candidate Recommendation (April 2008), http://www.w3.org/TR/WCAG20
5. Masip, L., Oliva, M., Granollers, T.: Hacia la semiautomatización de la evaluación heurística: Primer paso, categorización de heurísticas. In: Interacción 2010, September 7-10, pp. 359–368. Grupo Editorial Garceta, Valencia (2010)
6. Masip, L., Oliva, M., Granollers, T.: OPEN-HEREDEUX: open heuristic resource for designing and evaluating user experience. In: Campos, P., Graham, N., Jorge, J., Nunes, N., Palanque, P., Winckler, M. (eds.) INTERACT 2011, Part IV. LNCS, vol. 6949, pp. 418–421. Springer, Heidelberg (2011)

Model-Driven Development of Vocal User Interfaces

David Céspedes-Hernández, Juan Manuel González-Calleros,
Josefina Guerrero-García, and Liliana Rodríguez-Vizzuett

Computer Sciences Faculty, University of Puebla, Puebla, Mexico
{dcespedesh,juan.gonzalez,jguerrero,lilianarv}@cs.buap.mx

Abstract. There is lack of work addressing simply and extensively the development of vocal user interfaces considering at once the context of use: environment, user and platform. Several works have been published related to vocal user interface considered as a subset of bigger problems, such as: context awareness, multiplatform development, user-centred development, vocal user interface design, and multimodal development. It is normally the case to see that most design knowledge present in the literature assume vocal user interfaces as a subset of graphical user interfaces, called multimodal interaction, thus losing the nature of vocal interaction. The objective for this paper is to propose a method to generate multiplatform vocal User Interfaces. A transformational approach is used for the method. A real life case study is used to validate our proposal.

Keywords: Human-Computer Interaction, Vocal User Interfaces, Model-Driven Development, User-Centred Development.

1 Introduction

Recently, the way in which interaction with information systems is done is changing, this, due to the advance on technologies and the evolution of user interests. The keyboard and the mouse are being replaced by natural interaction supported by Natural User Interfaces (NUI) [1]. A particular type of NUI is a vocal interaction based system, software designed with the goal of emulating the dialog of a human being with another one. Vocal User Interfaces' importance is emphasized when taking into account the context of use of a system, for example, when the user can not use a graphical modality based system because of visual impairment or just because the user is working on another task at the time.

Voice eXtensible Markup Language (VoiceXML) (http://www.w3.org/TR/voice xml20/) is the World Wide Web Consortium (W3C) standard format for human-computer vocal interaction applications. Its main utilities are speech tuning and recognition but it was also used in [3] due to its simple structure as a notation for allowing tailorability in applications through a series of XSL transformations. Microsoft Kinect® Software Development Kit (SDK) (http://msdn.microsoft .com/en-us/library/microsoft.kinect.aspx) provides a toolkit that along with other features, allows the use of a microphone array (of a Kinect® sensor) for identifying the audio

C. Collazos, A. Liborio, and C. Rusu (Eds.): CLIHC 2013, LNCS 8278, pp. 30–34, 2013.

source and integration with the API of Microsoft speech recognition (http://msdn. microsoft.com/en-us/library/system.speech.recognition). The main purpose of this paper is to elaborate a set of transformation rules for the components of the Vocal Concrete User Interface model proposed and described on [4], to propose a set of icons for representing several of its elements as well, and to validate the model, icons and rules by applying them into a real life case study.

2   State of the Art

The evolution of interactive systems reached a point where today's research is centred in the development of NUIs, this is evident from the observation of the tools and new technologies that are offered in the market. This obeys at least two reasons, first the technological advancement that allows processing large amounts of data resulting on quick processing of natural interaction data, and second, that when well designed, the nature of this techniques makes the interaction easier and more intuitive.

Speech is the most commonly used communication method by the human being, as consequence of this, voice synthesis for giving information as output to the user and speech recognition to gather user inputs are a couple of highly developed knowledge areas. In [5], it is explained that the design of vocal UIs, involves several research fields, including human-computer interaction, psychoacoustics, digital signal processing and information visualization, raising the problem of a unified design process. The direct antecedent to this project is MultimodaliXML [6] where the objective of applying a set of XSL transformations over the specification of a concrete vocal UI model is mentioned, but it did not reach the implementation phase. There are also other projects like [7] and [8] that were implemented and support vocal interaction but not as its main objective and just for VoiceXML code generation, not allowing to develop interfaces for platforms that do not support that language.

3   Method

The meta-model proposed in [4] provides a full definition of components necessary to develop vocal interaction systems. The model is validated in the same paper by describing examples in which tasks of a system are defined in terms of the model. In this section, a set of icons is proposed to represent elements of the model in order to provide a graphical representation to the model concepts so they can be used in a software tool that supports the development of vocal UIs.

3.1   Graphical Representations for the Vocal Concrete UI Model Components

The design of several devices includes a representation for input and output sounds, as de facto standard, audio inputs are presented as a microphone and sound outputs are represented as speakers. This feature is respected and followed in the design of the representations for the vocal concrete UI model components. Since the other components to be represented act as containers, they can be represented as the combination of inputs and outputs to clearly show what they contain, for instance, a *vocalConfirmation* component involves an *output* and an *input* so it is represented as a box that contains an output icon plus an input icon. The same analysis is applied to every container in the model. The proposed icons for every component are shown in Fig. 1.

3.2   Transformation Rules for the Vocal Concrete UI Model Components

In order to develop vocal UIs starting from the definition of a system in terms of the vocal concrete UI model, a set of rules shown in Table 1 is provided.

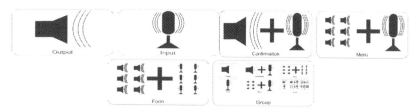

Fig. 1. Proposed icons for representing elements of the Vocal concrete UI Model

Once the rules are proposed and the graphic representation is provided, it is possible to design software with vocal interaction in terms of the model and to transform it into code specific to a platform. In the next section, a real life case study is made in order to validate the rules and representation proposed in this paper.

Table 1. Some of the Transformation rules for the vocal interaction meta-model

Component of the meta-model	VoiceXML Code
Output	<prompt> Text for the output </prompt>
Input	<field name = "identifier"> [Grammar] </field>
vocalMenu	<field name = "menuName"> <prompt> Instructions and menu options </prompt> [Grammar] </field>
vocalForm	<form id= "identifier"> [Content of the form] </form>
vocalGroup	<vxml version="versionNumber"></vxml>

4   Validation

As validation for the work described in this paper, an example is modeled in terms of the Vocal concrete UI Model from [4] but using the graphic representation proposed in the previous section. Later, VoiceXML code is generated using the transformation rules mentioned in the previous section too. The example proposed for the validation consists on a library information system that only reports if a title is in stock. First, the system should ask the user for an input. Then, once the user has provided the input, the system realizes a confirmation to check if it did understand correctly what the user said. Finally, if the user confirms the input, the system has to tell the user that the book is available. Fig. 2. Shows the model for the example described in terms of the vocal concrete UI model using the graphic representation proposed in this paper. The next step of the validation consists on using the rules proposed to generate VoiceXML code in order to provide the implementation for the designed system. The rules to be used in this particular case, area those for vocalGroup, vocalForm, Output, Input and vocalConfirmation. The order in which every component has to be transformed to code is determined by the arrows in the diagram from top to bottom and form left to right.

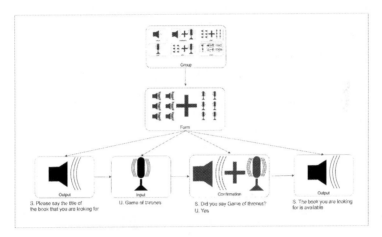

Fig. 2. Model for the book consult example

5   Conclusions and Future Work

In this paper, the Vocal concrete model proposed in [4] is supported by giving to its components a set of graphic representations in order to easily model vocal interaction systems on its terms. Also, a set or rules is presented for generating VoiceXML code from the model of a system. The future work consists on defining a set of rules for other vocal interaction supportive languages. Once these sets of rules are elaborated, a software tool must be implemented to support the full method and for realizing a validation process in which designers or non-expert vocal developers are asked to implement systems using that tool.

References

1. Câmara, A.: Natural User Interfaces. In: Campos, P., Graham, N., Jorge, J., Nunes, N., Palanque, P., Winckler, M. (eds.) INTERACT 2011, Part I. LNCS, vol. 6946, p. 1. Springer, Heidelberg (2011)

2. Steinberg, G.: Natural User Interfaces. University of Auckland (2012),
   https://cs.auckland.ac.nz/courses/compsci705s1c/exams/
   SeminarReports/natural_user_interfaces_gste097.pdf
3. Davis, V., Gray, J., Jones, J.: Generative Approaches for Application Tailoring of Mobile Devices. In: 43rd ACM Southeast Conference, USA (2005)
4. Céspedes-Hernández, D., González-Calleros, J., Guerrero-García, J., et al.: Methodology for the Development of Vocal User Interfaces. In: MexIHC 2012 Proceed-ings of the 4th Mexican Conference on Human-Computer Interaction (2012)
5. Daudé, S., Nigay, L.: Design process for auditory interfaces. In: Proceedings of the 2003 International Conference on Auditory Display, Boston, MA, USA (2003)
6. Stanciulescu, A.: A Methodology for Developing Multimodal User Interfaces of Information Systems. Ph.D. thesis, Université catholique de Louvain, Louvain, Belgique (2008)
7. Paternò, F., Giammarino, F.: Authoring Interfaces with Combined Use of Graph-ics and Voice for both Stationary and Mobile Devices. In: Proceedings of the Working Conference on Advanced Visual Interfaces, USA (2006)
8. Paternò, F., Sisti, C.: Deriving Vocal Interfaces from Logical Descriptions in Multi-Device Authoring Environments. In: Benatallah, B., Casati, F., Kappel, G., Rossi, G. (eds.) ICWE 2010. LNCS, vol. 6189, pp. 204–217. Springer, Heidelberg (2010)

Personalized Interactive Learning Solutions Support - IGUAL

Ion Mierlus Mazilu[1] and Esa Kujansuu[2]

[1] Department of Mathematics and Computer Science,
Technical University of Civil Engineering, Bucharest, Romania
mmi@mail.utcb.ro
[2] ICT and Software Engineering Department,
Tampere University of Applied Sciences, Tampere, Finland
esa.kujansuu@tamk.fi

Abstract. The overall objective of this paper is to present a IGUAL project solution used to improve the accessibility of higher education in Latin America for students from public schools. There is a measurable gap between the quality of education between private and public schools in most Latin American countries. This project will propose innovative, contextualised solutions, based on proved learning technologies, to help students with a public school background to rapidly close the gap and compensate for handicaps in their basic education. The specific objective of this project is to create and validate innovative and contextualised solutions to reduce the knowledge and skill gap between private- and public-educated students. These solutions will help the student to acquire new knowledge and skills, providing individually directed support based on the particular background and profile of the student. And also have the potential to be used by all students in the Latin American.

Keywords: education, learning software, pedagogical methodologies, interactiv learning materials.

1 Introduction

The low quality of primary and secondary education in most Latin America countries is a well-known problem. As a response to this reality, the private educational market has been steadily growing in those countries. These private schools, in general, offer a higher quality and personalized education for the students that can afford it. The main selling point of these institutions is access to better resources: better teachers, technologies, materials and pedagogical methods. This difference in education quality creates a problem once students from public schools reach university. The public schooled students have a strong handicap in their performance in a demanding and fast pace environment where professors are more concerned with the delivery of knowledge to large audiences than with catering to the specific needs of each student. This problem is aggravated by the fact that the great majority of public schooled students belong to low-income families. All the problems that arise from this social

C. Collazos, A. Liborio, and C. Rusu (Eds.): CLIHC 2013, LNCS 8278, pp. 35–38, 2013.
© Springer International Publishing Switzerland 2013

status in Latin America (need to work at an early age, economical difficulties, etc) also conspire to reduce the probabilities of success of these students. In this light, it is not just understandable, but to be expected, that the private schooled students out-performed their public schooled peers and gain better opportunities at the labor market.

The unequal primary and secondary education system in Latin America contribute to the inflexibility of the social mobility. Students that could afford private primary and secondary education have much better opportunities to have access to high quality universities and to complete successfully their studies. On the other hand, students that due to their socio-economical status only had access to public education have, statistically, a lower chance to enter universities and to obtain a professional degree. This has a negative impact on the competitiveness of Latin American countries, as only the middle- and high-income segments are fully contributing to the pool of specialized workforce while the talent and potential is uniformly distributed among the whole population. While scholarships and subsidized or free higher education could help to overcome the economic problems of low-income students, the lack of an adequate primary and secondary education has not been directly addressed in the region. While improving public basic education is the ideal solution to the problem, changing current educational structures have proved to be a long-term and difficult project for any country.

2   Analysis Research

The main problem to be addressed during this project is the increased level of difficulty that public schooled students confront during their university studies, compared with their private schooled counterparts. This difficulty results in lower performance and a higher level of dropout. The only way to deal with such great individual differences in the students will be to personalize the learning process for each student according to their current status and capabilities. A feasible and scalable alternative to personalize the learning experience of the students is to use learning technologies to create automated solutions to follow the students during their learning process, identify areas or skills that they require but are lacking, recommend them appropriate content from learning material repositories and guide them trough learning paths adapted to their individual needs. This project will generate learning solutions (combination of e-learning software, pedagogical methodologies and learning materials) to facilitate the assimilation of new knowledge and the development of new skills even when the student has deficient background knowledge and/or under-developed required skills. These learning technologies, initially developed in Europe, could be adapted to provide several "helpers" or "automated tutors" for each learner. Even if not perfect when compared with human tutors, these technologies could help disadvantaged students to receive the extra support needed to overcome the gap with their peers.

3   Proposed Solutions

The solutions that this project will provide have the potential to be used by all students in the Latin American Universities to support their learning process. However, it is expected that the public schooled students will be the main beneficiaries from these innovations, as they will enable those students to overcome the disparity in knowledge and skills with privately schooled students. In the context of the validation studies, at least 2000 students will be directly or indirectly involved in the different Latin American Universities that partner in this project by participating the pilot courses. If the project proves to generate positive results, this experience will be repeated each academic year and expanded to other knowledge areas apart from the pilot. The local companies are also final beneficiaries of this project. They can increase their competitiveness with new well-educated graduated professionals as their employees. Another final beneficiary group is the Latin American countries, as the know-how increases in the companies the competitiveness of the countries also increases.

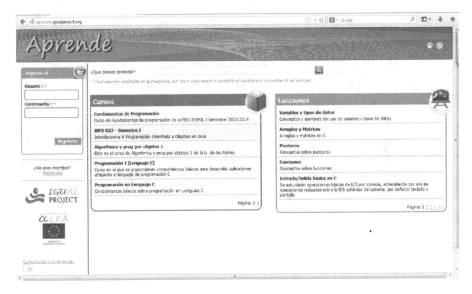

Fig. 1. http://aprende.igualproject.org/

This project supports the adoption of innovative learning technologies to solve current pressing issues will help in the modernization of Latin American HEIs. Moreover, these technologies are aimed to the most disadvantaged group of students and will be implemented in two of the poorest countries in the region and also the creation of the solutions will require an intense exchange of ideas, learning materials, tools, methodologies and results in a level not seen before in the region in the area of learning technologies. The IGUAL Project proposes the use of innovative learning technologies to help university students from public schools to bridge the knowledge and skill gap with their private schooled counterparts.

4   Conclusion

The overall objective of this project is to improve the accessibility of higher education in Latin America for students from public schools. There is a measurable gap between the quality of education between private and public schools in most Latin American countries. This gap has an immediate impact on the level of accessibility to higher education for each one of those groups. Students that come from the public schools have a lower probability to enter higher education institutions and also an even lower probability to finish successfully their studies. While there are several factors that are responsible for these results (need to work while studying, lower expectations), the knowledge and skill gap between private- and public-educated students is a key problem that aggravates the others. This project will propose innovative, contextualized solutions, based on proved learning technologies, to help students with a public school background to rapidly close the gap and compensate for handicaps in their basic education.

The specific objective of this project is to create and validate innovative and contextualized solutions to reduce the knowledge and skill gap between private- and public-educated students. These solutions will help the student to acquire new knowledge and skills, providing individually directed support based on the particular background and profile of the student. The personalized learning solutions will early detect problems with the students' knowledge and skill background, suggest students to review topics that were not being well covered in their basic studies and to recommend the student with activities to improve the level of their under-developed skills. Students with the public school background can use these tools to cope with deficiencies in their previous studies and to be up to par with their peers from private schools. The application area will be introductory computer programming, where the gap between public and private basic education is the widest due to the limited access to technological resources in public schools.

Acknowledgment. This work was supported by supported by European Union's ALFA programme an initiative of EuropeAid, IGUAL Project DCI-ALA/19.09.01/10/21526/245-315/ALFAHI (2010)123.

The authors thanks for assistance in the preparation of this paper to IGUAL team.

References

1. Ochoa, X., Cechinel, C., Jiménez, C., Arévalo, C., Araya, E., da Silva Camargo, S., Camerini, C., Chiluiza, K., Alvarez, L., Morales, J.: Need Analysis of the Students in Programming Courses in Latin America, http://www.igualproject.org
2. The IGUAL initiative: Innovation for Equality in Latin American Universities, http://www.igualproject.org

Closing the Gap between the Motivation of Users and the Design Requirements for Social Sites

Elizabeth Sucupira Furtado and Vasco Furtado

University of Fortaleza (Unifor), Fortaleza, CE - Brazil
{Elizabet,Vasco}@unifor.br

Abstract. The goal of this paper is to propose an extended format for describing interaction pattern making it an important artifact to associate aspects regarding the user's motivation with interaction solutions to design Social Systems (SS). 19 patterns, which were created, modeled in Semantic Media Wiki and applied in a case study, led designers to understand what motivates people to social involvement, and not just focuses on meeting the design requirements.

Keywords: Design Patterns, Social Systems, Users' motivation, Online Communities.

1   Introduction

Historically, when developing software, particularly during requirements analysis, the user's motivation was not a matter of concern. The most common context was that such motivation existed due to external factors (e.g., performing one's work duties). With the popularization of SS, one must not fail to understand what motivates people as members of communities. Requirements engineers and designers found themselves increasingly pressured to "socialize" their systems, often without understanding what motivates people toward social involvement, and consequently their engagement in interactions with SS.

Even though the motivation to join communities is the object of study in various fields, such as social psychology [1], HCI [2], appropriation of the results obtained from these studies by software engineering is not trivial. What we see today, in practice, are very few methods and instruments for professionals to understand people's motivations within communities (in short, social motivations), which prevents them from perceiving opportunities to create interactive patterns that facilitate this social involvement.

Our research contribution aims to provide a way of listing forms of motivation of members of an online community, from the viewpoint of studies conducted by Social Psychology, by specifying interaction solutions for a SS. We extend the original formalism of interaction patterns [3] in order to include specifications on two levels of abstraction. At the conceptual level, the specifications refer to users' intentions (such as their feelings, needs, etc.) for interacting in a SS, and to the motivational factors related to such intentions. At the design level, the specifications present a solution for

C. Collazos, A. Liborio, and C. Rusu (Eds.): CLIHC 2013, LNCS 8278, pp. 39–46, 2013.
© Springer International Publishing Switzerland 2013

interaction; an analysis is made in association with such solution regarding its implication on user behavior, describing hazards that may compromise the use of the SS by users. Based on this specification, we can start building a base of interaction patterns, available in a wiki environment [4] in order to be accessed and updated, if necessary, by the academic and industrial community.

An evaluation of the proposed patterns was conducted with a comparison between designers who used the patterns and those who used interaction patterns commonly found in HCI literature, during a phase of requirement engineering of a SS. The results indicated differences vis-à-vis the design solutions and the understanding of designers about features that motivate the user to interact with the SS.

The structure of this paper is as follows: after a brief review of the literature about studies that bridge the social sciences and the design of SS interaction, we present concepts about social motivation. Then we address the formalism of the proposed pattern, as well as the results achieved in defining and evaluating the defined patterns.

2   Background Knowledge

The need to consider the social motivations in SS development has been addressed by researchers in the fields of domain engineering and usability engineering.

Domain engineering is aimed at identifying and modeling common characteristics and variables in applications of a given domain that allow one to develop domain models as patterns, classes and requirements of the domain [5]. For the SS domain, the most commonly produced artifacts are the requirements for the social practices, specializing in: (i) personal values, relating to feelings such as trust, and ethics; (ii) mode of communication, referring to user participation; and (iii) mode of treating content, such as data sharing, maintaining one's privacy, copyrights, etc. Usability Engineering considers usability from the start of a project primarily making use of guidelines, pre-patterns [6] and interaction patterns [7] for defining design solutions. The concept of new usability [8] extends the concept of usability, which focuses on the efficiency and effectiveness of performing the tasks that the user does in the system, to equally consider social practices (such as Collaboration, Communication, Ethics, Added Value) and those related to the user's experience with the SS (such as Security, Trustworthiness and Privacy). It is quite similar to the UX facet concept.

We conducted an extensive analysis of studies related to these fields of engineering and that, due to space restrictions, we will not detail. Particularly what motivates us in this article is to explore the fact that none of these works presents artifacts that make associations between design solutions and motivational factors, which may be useful to specify requirements related to the new usability. At best, these studies propose to consider requirements of W3C's 3C model (Collaboration, Communication and Co-operation) [9], but make no reference to any motivational factor. Some proposals advance in this matter [10-15], but remain at the model level and require tremendous effort of interpretation and/or learning by designers. For example, there is no base of cases or patterns (equivalent to Welie's proposal, for interface patterns in general [16]) readily accessible to designers and in which the interactive objects are related to the motivational aspects of social participation.

3   Social Motivation: A Perspective from the Social Psychology

People's motivation to participate in communities is explained by two base theories: Identity-based Community (IC) and Interpersonal Relations-based Community (IRC). The first one explains people's motivation from the perspective of attachment they begin to feel with the community, while the second one refers primarily to the connections made with people who are part of the community. These theories were explored in a study conducted by [12], which describes the main causes that lead people to join communities. The three main causes that lead the user to identify with the community (attachment to an IC) are the following:

- Social categorization refers to the fact that people naturally create an identity with a group that nominates them and therefore start to feel committed thereto. This can be a group of fans of a particular football team;
- Attachment by interdependence occurs when one needs to solve tasks that require cooperation, where a common purpose is shared, such as attaining a high score in the group. Example includes crowdsourcing initiatives, where people contribute to the achievement of a particular goal; and
- Comparisons between groups are also the driving force that motivates people since by exercising this sense of comparison, people wind up identifying more with the community to which they belong. This cause explains the use gamification strategies as a way to motivate user participation.

The three main causes that lead users to have interpersonal relationships with a community (attachment to an IRC) are the following:

- Social interactions allow people to meet and establish trusting relationships. The more they interact, the more likely people will establish a relationship. Attachment also increases when people begin to feel they are jointly present in a virtual space. This is a feeling of co-presence that fills people;
- Opportunity to exchange information, usually intimate, about oneself and others is a cause and consequence of interpersonal bonding. Sharing people's habits, finding the times that they log on to chats, for example, is a way to create these links. Even people who do not interact seem to share bonds when personal information is shared; and
- People are fond of others who are similar to them in terms of preferences, attitudes and values, and therefore interactions and relationships between similar people are enhanced. Similarity can create not only a bond relationship, but also a common identity to the group that involves the "similar" members.

Ren et al., [12] conclude that, depending on the type of bond that one has with the group, one's behavior may vary in relation to several aspects, like: the content of the discussion, the way responsibility is distributed, adaptation to the rules of the group, stance regarding new members, reciprocity, and the implications as to the robustness of the community. Off-topic content is much more welcome by those participating in an IRC, while users of an IC prefer to dialogue on topics related to the purpose of the community. Members of an IC-type community tend to take on more responsibility for absent or non-engaged people, while those in an IRC-type community tend to try to hold accountable culprits or those with little responsibility. The stance of new users and the stance of existing members toward new uses also varies. New users have more difficulty to mingle in an IRC, because the members – since they already know one another – are more difficult to approach (the popular "breaking into the clique").

In an IRC, members have more reciprocity to the community in general. There is a sense of altruism that is more easily promoted. Ultimately, the robustness of the community also varies depending on the type of community. Those who feel attached to the group are less susceptible to changes in membership. However, for those who are more attached to the group, if the friends leave, the community loses interest.

We believe it is important designers to know the implication of these aspects for "user behavior" in order to design solutions for SS and to develop a rationale for their solutions based on the users' attachment feeling. So we consider patterns, as the artifact of linking users' motivation to design solutions for SS.

4   Patterns for Representing Social Motivation

The basic motivations considered for the choice of patterns, also as the artifact of reusing specifications in this article, were as follows: can be used for communication, since a pattern is described in natural language; serve as a bridge among specifications of different levels, favoring the assessment of requirements with the use of prototypes; and serve to capture knowledge from previous projects, enhancing the analysis and design of a user experience adapted to one's intentions [13].

The patterns proposed in this article are innovative because they support **the association of motivational factors with reusable solutions for SS design** through a new format for defining design patterns. The format of the proposed patterns is inspired by the original format defined by [3], but fields have been added that allow us to represent matters related to motivational factors. Table 1 describes the fields introduced.

Table 1. New information existing in the proposed patterns

New Information	Description
Designer's assumption	what the designer believes to be the user's motivations, feelings and needs;
Rationale for the assumption	what led the designer to have the assumption of what the user's motivations are vis-à-vis the characteristics of the community the user belongs to;
Solution	the design specification or interaction in an SS in terms of conceptual interface components for specific user activities;
Impact on user behavior	what the implication is for "user behavior," when using the pattern in the design of an SS. Some aspects are in section 3;
Hazards	the factors or risks that the application of a pattern may cause, relating to negative feelings of users.

Figure 1 shows an example of the pattern *Awareness of entry of new users*, which aims to describe ways to show the user – in one's peripheral vision – that new users are joining the SS, available in [4].

Awareness of entry of new users

Synopsis: Introduce new users into the community.

Designer's Assumption: Users will like to know the success of their network. New users keep a community alive because they rejuvenate it and keep it vibrant.

Rationale for the Assumption:

If IRC: Display of new users reaches users in their motivations to meet new people:
a) social interactions: knowing that your friends are also coming to the community
b) interpersonal similarity: feeling that your friends share your likes   and you can talk to them

If IC: Display of new users aims to foster interaction between people who do not necessarily know one another, but may have common goals:
a) Identity: Feeling increasingly proud to be part of a community that is growing
b) Group interdependence: finding new partners to take part in actions that are common to the community and in shared activities
c) Comparison between groups: compare one's community with other communities and understanding their growth dynamics

Context: The perception that new users arrive in the community induces a subsequent interaction so that they can help reach its goals and fulfill its motivations.

Solution: A space for interaction in which the user perceives who the new users are. It is also advisable to send welcome messages, making entry into the community more enjoyable.

Impact on user behavior in IRC

a) Receptivity: A user of this community should know when friends from other social networks join the community. However, if it is not a friend of theirs, users tend to dislike it when the network grows too much.

Impact on user behavior in IC

a) Receptivity: Old users can identify potential companions in solving tasks or strengthening community identity. Whenever possible, displaying new users should be enriched with information to establish relationships as their own expertise and specific preferences.
b) Commitment: The participating user becomes aware that people are coming in, which are likely to be passive at first, and it's important to engage them in discussions. The display of new users should be enriched with information that can help them get involved in the discussions;
c) Rules: Old users, especially more experienced users, usually see this as a way to introduce newcomers to the community's rules.

Hazards: Displaying the entry of new users can be frustrating, if it occurs at a low cadence. It can demonstrate a low dynamic of the community and – rather than motivating – can discourage users. On the other hand, in IRC, users are resilient and usually do not like to know that the network is not robust, with constantly changing people.

Examples: Facebook indicates when invited friends enter the community and asks the user to receive them and to indicate new friends. The Virtual Cheering Squad shows new fans who join the community enabling them to make new friends and assist in the search to get points in the contest to become the squad leader.

Fig. 1. New user entry awareness pattern

5   Evaluation

With the aim to evaluate the patterns created, we conducted two system-design sessions with six young designers (graduate students). All of the designers had knowledge about new usability [8], characteristics of community (cf. Section 3) and interface design. We focused on Requirements Engineering, which occurs in the early stages of developing a SS and supports the elaboration of interaction design solutions. The domain chosen was that of an airline check-in system, which we call social check-in. In it, a community of passengers is formed (it is an IRC) and the following (design) requirements were specified: opportunity to interact with travelers who follow the same segment, messaging among travelers in general, choice of seats close to acquaintances, different ways to receive the boarding pass, sharing seat selections, data retrieval of friends on Facebook, and invitation to join the community.

In the first session, we suggested that each designer elaborate an activity stream for one scenario that described a happy day situation, involving the requirements. They should follow the Model of user EXperience model (MEX) [17], because it captures the essence of a scenario of experience (as the expected behavior of the target-users and their activities in the system). In the second session, they would carry out the designs the system's interface. Half of the designers (group 1) were offered the interaction patterns proposed herein, which suggest considering the social motivations during the early stages of design. The other half (group 2) used interface patterns without the new proposal and at available http://mobile-patterns.

Our evaluation was aimed at validating the assumption that by using the patterns that take into account the motivational aspect of the users, the system design would be richer, because it would consider more features aimed at entertaining the community. The very prototypes of the interfaces were the products used in our analyses.

Firstly, the analysis showed that designers of group 1 created several functionalities focused on social engagement more than the others. For example, the three designers who used the extended pattern proposed a feature for creating user groups with common interests, while only one of the group 2 did likewise. Other features proposed by all the designers of the group 1, are cited below, with the respective patterns used: show who was online on the network (co-presence awareness pattern); show newcomers (new user awareness pattern); describe the purpose of the network so as to lead members to interact (social goal presentation pattern); show the number of members of each group created (group success pattern); and, especially, create a notification area that lets the user know what's happening in the community at all times (group interactions awareness pattern). None of those that were based only on the mobile-patterns proposed these or similar functionalities.

In addition to this qualitative analysis of the prototypes, we conducted a questionnaire to be answered by the designers in order to assess their perception about the motivational aspects to be taken into account in their designs. Two questions were asked: i) Which features of your system exist to form a community with common interests involving air travel?; and ii) Which features did you design to motivate the user to interact and attain the social goal?. The responses helped us to capture the degree to which usage of such patterns induced in the designers a reflection of the importance of creating appropriate strategies for the motivation of community members. The answers to the first question showed evidence that the designers of group 1 perceived the motivational factors (such as attraction by similarity), which were important for the type of community being formed in the SS. The designers of group 2 – even though their designs presented users with travel in common, failed to perceive what attracts them, aside from the fact that they were going to travel together. The main focus of their designs was to allow users to choose people to sit next to them. Below are a few excerpts of the responses of group 1: "In the social check-in, there's an IRC formed around groups"; "the users attract one another mainly because they're flying the same segments"; "there's interaction among users who have the same likes, the division of groups by common travel." The main focus of their designs was one whereby the user can create groups (network organization pattern), invite friends to join (social network increment pattern, social objective presentation pattern), and discuss issues relating to flight segments (group interaction awareness pattern, co-presence awareness pattern, new user awareness pattern).

However, when evaluating the second question, we find the response of one of the designers – the same one who proposed the creation of groups of users with common interests – that shows that he might have reflected on the motivational issues. He replied: The user can become more 'popular' by sharing his/her flight information on Facebook. Despite this response, he was unable to propose feasible solutions for this in his design, as already analyzed in his prototypes. One extended pattern that we propose could be used to reflect upon the implications for motivation would be the ranking of participation. The other two said that allowing interaction with people (whether acquaintances or not) prior to the flight was a form of motivation. The designers of group 1 referred primarily to the notifications in the social network Facebook, such as messages regarding how many trips your friend took, reminders of how to use the social interaction options (help pattern), descriptions on how the community works (social objective presentation pattern), notification of sharing of check-ins (group interaction awareness pattern).

Specifically for group 1, we asked yet another question: Do you think that the concepts of extended patterns helped you to make the design? They agreed, with the following justifications: "It helped in consideration of why to include that pattern in the solution, for the type of community that we wanted to create. But some concepts thought were modified when implemented due to limitations of the device"; "Through the pattern, I understood the importance that should be given to motivational factors. The justifications helped to identify situations in which the patterns would be used"; and "The patterns defined were extremely helpful and really facilitated the design of the interface for this SS. The assumption associated with the pattern shows a user's behavior that would be expected for that situation, facilitating the simulation of a scenario to be included in future system usability testing."

6   Final Considerations

This paper brings up a discussion about the importance in using patterns, and in this case, through the new format, which associates motivational factors with interaction solutions. The patterns are being used in a methodology that supports the development of interaction design solutions based on a model, which captures the essence of the association made between the users' behavior and their activities and the specifications of the proposed patterns.

It is important to highlight two points: first, new patterns can be described focusing on other motivational factors and concepts available in social science literature. Second point is related to the evaluation results and process. This qualitative analysis verified that the designed functions of engagement were related to proposed patterns (their effectiveness). In addition, designers that used them, developed a rationale to explain their decisions, based on the concepts adopted in this text. The supposition raised confirms the need for the proposed patterns. No consideration was made as to the time that the study participants began to use the patterns (their efficiency). In the future, quantitative studies should be conducted with a larger sample. Due to these facts, there is limitation in the results, which are not considered conclusive.

References

1. Beenen, G., et al.: Using Social Psychology to Motivate Contributions to Online Communities. Human-Computer Interaction Institute. Paper 88, 1–12, 23 (2012),
   `http://repository.cmu.edu/hcii/88`
2. Pereira, R., et al.: A Discussion on Social Software: Concept, Building Blocks and Challenges. International Journal for Infonomics (IJI) 3(4) (2010)
3. Alexander, C., et al.: A Pattern Language. Oxford University Press, NY (1977)
4. Motivational Patterns wiki system,
   `http://200.19.188.105/prepadroes/index.php/`
   `P%C3%A1gina_principal`
5. Kang, K.C., Cohen, S.G., Hess, J.A., Novak, W.E., Peterson, A.S.: Feature-Oriented Domain Analysis(FODA) Feasibility Study. CMU/SEI (1990)
6. Saponas, T.S., et al.: The impact of pre-patterns on the design of digital home applications. In: DIS. ACM, University Park (2006)
7. Chung, E.S., et al.: Development and Evaluation of Emerging Design Patterns for Ubiquitous Computing. In: DIS. ACM, Cambridge (2004)
8. Thomas, P., Macredie, R.D.: Introduction to the new usability. ACM Transaction on Computer Human Interaction 9(2), 69–73 (2002)
9. De Oliveria, L., Gerosa, M.: A Domain Engineering forContent Sharing Collaborative Features. In: Webmedia. Brazilian Computer Society (2012)
10. Piccolo, L., Baranauskas, C.: Basis and prospects of motivation informing design: Requirements for situated eco-feedback technology. In: IHC. SBC, Cuiabá (2012)
11. Lampe, C., Wash, R., Velasquez, A., Ozkaya, E.: Motivations to participate in Online Communities. In: CHI 2010 (2010)
12. Ren, Y., Kraut, R., Kiesler, S.: Applying Common Identity and Bond Theory to Design of Online Communities. Organization Studies 28, 377 (2007)
13. Gurzick, D., Lutters, W.: Towards a Design Theory for Online Communities. In: Desrist. ACM (2009)
14. Mohamed, N., Ahmad, H.: Information privacy concerns, antecedents and privacy measure use in social networking sites: Evidence from Malaysia. Computers in Human Behavior 28 (2012)
15. Zhou, T.: Understanding online community user participation: A social influence perspective. Internet Research 21 (2011)
16. Welie, M., Troetteber, H.: Interaction Patterns in User Interfaces (2000)
17. De Carvalho, C.: MEX experience boards: A set of agile tools for user experience design. In: IHC. Brazilian Computer Society (2010)

Formative Evaluation for Complex Interactive Systems

Chris Roast and Elizabeth Uruchurtu

Culture, Communication and Computing Research Institute
Sheffield Hallam University
153 Arundel Street,
Sheffield, S1 2NU, UK
{c.r.roast,e.uruchurtu}@shu.ac.uk

Abstract. This paper reports upon the design and use of a lightweight evaluation method, especially designed to examine complex interactive systems. The approach is illustrated through a case study involving an interactive tool designed to help enable users examine large scale data arising from authentication activity in higher education institutes. The evaluation approach illustrated is to enable the lightweight assessment of usability issues within complex interactive systems and identifying opportunities for significant design improvements. Specifically we argue that this method benefits from capturing key generic factors that underpin the effectiveness of tools for working with complex data. The paper concludes by reflecting upon the effectiveness of the lightweight structured assessment approach and how it supports to formative evaluation.

Keywords: Evaluation, Cognitive Dimensions, Complex Data, Information Retrieval, Innovation.

1   Introduction

Approaches to evaluating interactive systems are wide and varied and can be judged in terms of the value of the outputs that they provide and the effort required in obtaining the outputs. The technique illustrated in this paper is formative and lightweight in character, and also specifically suited to complex interaction. Our technique is derived from a framework that was developed to capture human factors evident but rarely touched-on with conventional techniques. The type of complex interaction of interest here are those where user tasks involve working with notations, or languages, in order to achieve a desired effect. Specific complexity arises when the notation has tokens with powerful indirect meanings. A simple example would be an electronic calendar that supports recurring appointments, the means of defining a recurring appointment introduces significant new possibilities that users should be conscious of. In general, we judge the subsequent complexity to result in "programming-like" activities - such as, finding and fixing mistakes with a recurring calendar appointment. Thus, the effective use of such systems is not only reliant upon the appropriateness of the mechanisms available to manipulate the notation, but also upon the user interpretation of how the system might process the notation. This often results in an intrinsically

C. Collazos, A. Liborio, and C. Rusu (Eds.): CLIHC 2013, LNCS 8278, pp. 47–54, 2013.

indirect manipulation. Programming development environments provide obvious examples of this type of complexity in interaction. But similar complexities can be found in more mundane systems such as: calendars and online booking systems.

The case study in this paper is a powerful authentication monitoring tool. It supports the articulation and execution of complex searches of large datasets and as such is "programming-like". In brief, the tool's control panel allows the selection of search templates, specification of filters and parameters and the specification of the data to be output as a graph (illustrated in figure 3). This was viewed embodying the indirect manipulation common with complex interactive system. In addition, a brief tutorial about the tool highlighted how a range of configurable filters could be used to generate user defined views of data.

2   The Analytic Framework

The evaluation approach is a collaborative lightweight method motivated by concepts taken from the Cognitive Dimensions framework [4]. The framework has been the focus of considerable research interest, its potential as a tool for evaluation has been explored with a number of approaches [5]. One example of a framework dimension is "Secondary Notation" - this focuses upon how a system may enable unstructured attributions to a notation (such as comments or highlighting). The framework has some similarities to the concepts of "design patterns" [3] and "ergonomic criteria" [12]. However its relevance for this research comes from its descriptive nature and its use in examining notational systems.

Authoritative sources for the framework show a diverse range of such dimensions grounded in concrete examples with informal definitions. Research into the dimensions framework has predominantly focused upon their adoption through the comprehensive and consistent use of the dimensions [1,2,9]. Hence, methods for assessing concepts such as "Secondary Notation" have been explored with the aim of providing an objective assessment of them. Although this is clearly valuable, these endeavours appear to have overlooked the fact that the illustrations of the dimensions also demonstrate insights into potential designs that help innovate design alternatives. So, in the case of "Secondary Notation" the different uses to which unstructured attributes might be put based upon examples and analogies can be insightful. Examples of the uses of "Secondary notation" include: a means of communication, a facility to improve presentation, as well as a technique for demonstrating expertise. While these points are worthy of evaluation, they in fact point interesting ways in which a notation might get used.

Hence, instead of treating the framework as a means of assessment, it also has the potential to promote innovative perspectives upon existing designs. The approach to formative evaluation described here follows this line of argument and thus places less priority on objective comprehensive assessment and more on the variety of ways, or modes, in which concepts found in the Cognitive Dimensions framework drive new ideas or insights.

2.1   The Tabular Framework

Our approach is to use a simple tabular format for engaging system developers, experts and end users in co-operative evaluation. The use of this has been reported [9,10] within the context of a tool for digital video post-production and publishing. The tabular approach is designed to encourage collaborative reflection and insight through focusing upon a relatively small number of key questions (derived from those in [1]).

What are the dominant / common ways in which these concepts are shown together or reached from one another?

from \ to	Specification(s)	Data set(s)	Publisher service(s)
Specification(s)	always / no. of clicks / not during ...	always / no. of clicks / not during ...	always / no. of clicks / not during ...
Data set(s)	always / no. of clicks / not during ...	always / no. of clicks / not during ...	always / no. of clicks / not during ...
Publisher service(s)	always / no. of clicks / not during ...	always / no. of clicks / not during ...	always / no. of clicks / not during ...

Fig. 1. An example tabular entry to examine how easy it is for the user to navigate information

Reflection and potential insights are encouraged by the tables presenting how, for a single question, it could be answered from a number of perspectives. For instance a single table will encourage users to respond to a question such as "How is A reached from B?" and also "How is B reached from A?". Figure 1 shows an example for this type of question in full using three alternative concepts from the case study and with indicative possible responses within the cells. Through this instrument users are encouraged to explore ideas that they may not normally consider. The use of just three alternative concepts keeps the approach tractable for collaborative assessment.

The three core concepts used are chosen to be ones central to effectively performing the work that a target system is aimed to support. Concepts are chosen to be relevant, high level and ideally encompass a number of potential conceptual "mismatches" as described in [2]. Space does not permit a detailed description of the selection of concepts. However, it is worth noting that the selected concepts serve as a familiar basis for analysis and reflection and as such their precise definition and consistent use is not central to facilitating.

2.2   The Facilitation

Operationally the tables are presented on paper to encourage ease of engagement and enable additional points to be easily recorded. While a subject may use the suggested response alternatives, they can just as easily respond in a manner that is more appropriate for their task and interest. For instance, they may even sketch thumbnail illustrations of what is implied by a specific cell. While the form of the process is relatively simple, the facilitator works with system users and/or experts to build their confidence in completing the tables and encouraging deeper reflection. Notes and marks on or beside the tables are encouraged to reflect and record any other opinions or views. The tables encourage users to make relative assessments within each table, discouraging default responses. In addition, the facilitator encourages the completion of the tables by asking for concrete illustrations or examples of particular judgements.

There are two general roles of the facilitator: to encourage reflection, and to record reflection. The facilitator's activity is to primarily work on the first of these and then ensure the second is provided by the participant.

2.3   Ideas and Insights

Having completed the table entries the facilitator and participant will have reflected upon the nature of the tool being examined and in doing so will be able to identify potential improvements. The value of employing the tables and their links to the Cognitive Dimensions framework is that the resulting observations are: (i) expressed in generic structural terms and not in terms of local corrections or "fixes"; (ii) the framework can provide insights into ways in which particular dimensions re-frame the system being examined. Earlier we provided an example of this when one considers "Secondary Notation" - once an annotation is pointed out as one way improving presentation and same possibility can be explored with the target system.   Overall for each table, alternatives and re-framings can be suggested and examined.

3   The Case Study Context

Our case study concerns the management of online resource authentication within educational institutes. Specifically the system examined supports the monitoring and assessment of subscription services in order to understand how services are used. It's development was supported by JISC and it is currently adopted by a number of UK universities[1]. The direct users are library staff and library managers who may need to review service uptake and, say, compare similar services. A specific example might be to identify whether computing students use the ACM Digital library (www.acm.org/dl) on a comparable basis to IEEE Explore (ieeexplore.ieee.org), or whether in terms of usage, say, Sciencedirect (www.sciencedirect.com) effectively subsumes both.   At face value this may not appear to be a particularly complex task, but the raw authentication data often hides subtle details. Some authentication events match one-to-one with accessing a publication, while others can be one-to-many, and on some occasions many-to-one. Such differences arise when each service chooses what authentication standards and policies they will use. In short, comparing service is a non-trivial exercise of interpreting mixed data sources. The case study tool is

---

[1] See JISC website:
http://www.jisc.ac.uk/whatwedo/programmes/aim/raptor.aspx

designed to help address some this complexity by integrating authentication event logs and to examine aggregate views of them over time.

3.1   The Tool's User Interface

The tool examined consists of three architectural components: a web front-end; an aggregator that collates and stores authentication data and performs searches; and, agents that send event logs to the aggregator. End users interact via the web front end which provides access to a "graphs" page. Figure 2 provides an illustration of this page, simplified to highlight the key structure and elements. On this page the user is able to build a search specifications, using a number of given types-of-search forms. Within each they are able to specify details such as: (i) the type of authentication protocol to examine; (ii) the date range of interest; (iii) the level of granularity of the resulting data; (iv) a series of filters that can be used to exclude authentication log items based upon characteristics of the log entries; and (v) a series of post-processors that determine alternative data presentations. In addition, users can choose to provide labels for data sets generated by a search and also names for the filters as they are applied. Having formulated a search in this manner, the data set can be generated with the click of an "update" button. The resulting data set is shown as a graph, with the options to access the same data in different formats and reports.

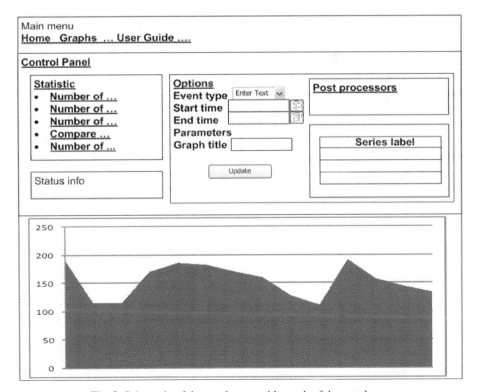

Fig. 2. Schematic of the graph page with graph of dataset shown

4   Applying the Framework

4.1   Initial User Interface Review

Intended primary and secondary users were part of the project team. The team collectively reviewed the system, trying out its functionality, led by an end user while verbally reporting to the team and responding to their comments or questions. The authors focused upon supporting the team in exploration and experimentation with the tool. The first author led this process while the other observed and recorded team reactions and comments. In a subsequent review of the notes, the observations were mapped to the tabular framework, with the questions from the framework acting as "leading questions" in interpreting observations. The review meeting discussed a wide range of tool and usage activity related concepts, and some generic high level usage models: (i) Finding data, (ii) Comparing data, (iii) Aggregating and interpreting data, and (iv) Mixed source assessment. These helped develop a common understanding of potential usage and direct the assessment of the tool. In particular the following three concepts were identified for the assessment of the tool:

- **Searches** Data set specifications for finding data or comparing data
- **Data** The results computed by the system in response to a data set specification
- **Services** Subscribed to services, how they appear in data sets and also their relative cost.

How easy is it to change or alter relationships between concepts?

	Publisher service(s)	Data set(s)
Specifica-tion(s)	**Hard - Some of the filtering parameters seem to allow this but they are unclear**	**The relationship is that of the search done ... the post processors and filters can narrow or broaden the relationship. Broadening is easier than narrowing.**
Data set(s)	**Hard - a data set is a product of one specific set of services at the point at which the data set was created.**	

Fig. 3. An example of tabular entries for the target system focusing upon ease of change

Figure 3 illustrates one of the tables produced from this review.

The subsequent analysis of the completed tables involved: (i) assessing how coherent relationships and how well the underlying factors were understood; and (ii) taking examples of the table relationships and re-examining them in terms of the alternatives suggested by the dimensions. These two processes drove further consultation with the team and allowed alternative design ideas to be examined. These improved our understanding of the authentication infra-structure and the standards used, while also helping examine different ways in which authentication data might be analysed, structured and managed.

4.2   Case Study Outcomes

Case study outcome can be expressed in terms of the concepts underpinning the tabular form. Here we summarise the outcomes focusing upon: abstraction and consistency.

Abstraction. Abstraction mechanisms largely concern the creation and management including explicit and implicit approaches. In our assessment: (i) while the specification of individual searches is supported by the tool, the notion of a search with a generic task oriented purpose is not. Hence, a search that might be conducted to form the basis of a monthly report, is not supported.

Our consideration of data focused upon the value of placing search results next to each other. For instance although it is valuable to compare data in across comparable timeframes (e.g. seeing Jan 2011 data next to Jan 2012 data), the tool only supports this if the timeframe are the same.

Consistency. Consistency as user interface principal is broadly accepted ([8,12]) while the specific meaning and merits are dismantled with ease ([6]). In the context of the tabular approach, consistency is captured by the lack of confusion between entities represented within a system. In the case of searches of the authentication data the potential to confuse searches is significant, the numerous of parameters, some of which are only shown via sub-dialogues, do not help. Hence, the differentiation between searches is very difficult.

In a related manner the potential for confusing data sets is very high as it is the user's responsibility to remember to provide a meaningful label for the data set when the data is generated.

5   Reflections and Conclusions

The evaluation of complex interactive systems following conventional approaches demands considerable effort and resource. Users and stakeholders are hard to access and as a consequence, evaluation tends to yield lists of issues in a similar vein to those generated from Heuristic evaluation [7]. While these issues can be collated, prioritised and addressed they rarely capture key structural factors that are especially important when examining complex interaction.

By contrast our tabular approach links together a rich framework for exploring complex interaction, with a relatively easy form of conducting analysis and developing formative outcomes from that analysis. We believe this approach is of particular value since it appears to be less resource intensive while also generating insights regarding potential design alternatives. All of the issues identified in the case study assessment have been accepted by the development team as requiring solutions, with some being directly addressed.

The tabular approach described offers a method focused upon innovative formative evaluation. With the growing use of powerful data intensive systems and the likelihood that non-expert users are expected to engage with them, ensuring that such "empowered" users can work effectively is of high importance.

Acknowledgements. The authors would like to acknowledge the JISC Access & Identity Management Programme for supporting this work and staff and students at the host institution for participating in user based assessments.

References

1. Blackwell, A.F., Green, T.R.G.: A Cognitive Dimensions questionnaire optimised for users. In: Blackwell, A.F., Bilotta, E. (eds.) Proceedings of the Twelfth Annual Meeting of the Psychology of Programming Interest Group, pp. 137–152 (2000)
2. Blandford, A., Green, T.R.G., Connell, I.: Formalising an understanding of user-system misfits. In: Bastide, R., Palanque, P., Roth, J. (eds.) DSV-IS 2004 and EHCI 2004. LNCS, vol. 3425, pp. 253–270. Springer, Heidelberg (2005)
3. Dearden, A.M., Finlay, J.: Pattern languages in HCI: A critical review. Human Computer Interaction 21(1), 49–102 (2006), doi:10.1207/s15327051hci2101_3
4. Green, T.R.G., Blackwell, A.F.: Design for usability using Cognitive Dimensions. Tutorial Session at British Computer Society Conference on Human Computer Interaction (HCI 1998) (1998), http://www.cl.cam.ac.uk/~afb21/CognitiveDimensions/CDtutorial.pdf
5. Green, T.R.G., Blandford, A.E., Church, L., Roast, C.R., Clarke, S.: Cognitive dimensions: Achievements, new directions, and open questions. Journal of Visual Languages and Computing 17(4), 328–365 (2006), doi:10.1016/j.jvlc.2006.04.004
6. Grudin, J.: The case against user interface consistency. CACM 32(10), 1164–1173 (1989), doi:10.1145/67933.67934
7. Nielsen, J., Molich, R.: Heuristic evaluation of user interfaces. In: Chew, J.C., Whiteside, J. (eds.) Proceedings of the SIGCHI Conference on Human Factors in Computing Systems (CHI 1990), pp. 249–256. ACM, New York (1990), doi:10.1145/97243.97281
8. Nielsen, J.: Usability Engineering. Academic Press, Boston (1993)
9. Roast, C., Khazaei, B.: An investigation into the validation of formalised cognitive dimensions. In: Doherty, G., Blandford, A. (eds.) DSVIS 2006. LNCS, vol. 4323, pp. 109–122. Springer, Heidelberg (2006)
10. Roast, C., Dearden, A., Uruchurtu, E.: Using and utilizing an innovative media development tool. In: Proceedings of the 10th Brazilian Symposium on Human Factors in Computing Systems and the 5th Latin American Confernce on Human-Computer Interaction, pp. 145–156 (2011)
11. Roast, C., Uruchurtu, E., Dearden, A.: The programming-like-analysis of an innovative media tool. In: Psychology of Programming Interest Group Annual Conference. University of York (2011)
12. Scapin, L.D., Bastien, C.J.M.: Ergonomic criteria for evaluating the ergonomic quality of interactive systems. Behaviour and Information Technology 16, 220–231 (1997)

Setting Usability iTV Heuristics in Open-HEREDEUX

Andrés Solano[1], Llúcia Masip[2], Toni Granollers[2], César A. Collazos[1],
Cristian Rusu[3], and José Luis Arciniegas[1]

[1] Universidad del Cauca (Colombia), Departamento de Sistemas, Grupo IDIS
{afsolano,ccollazo}@unicauca.edu.co
[2] Universidad de Lleida (España), Escuela Politécnica Superior, Grupo GRIHO
{lluciamaar,tonig}@diei.udl.cat
[3] Pontificia Universidad Católica de Valparaíso (Chile), Escuela de Ingeniería Informática,
Grupo UseCV
cristian.rusu@ucv.cl

Abstract. Usability evaluation is one of the most important stages inside the
user-centered design. Heuristic evaluation is a highlighted method due to its
great capability to detect a high number of usability issues, its low cost and
simplicity; however, the search and selection of the best heuristics is probably
the most difficult task of the methodology. This paper describes the process carried out to enter a set of heuristics (for interactive digital television environments) to Open Repository of the Open-HEREDEUX, which aims to minimize
the effort to select the most suitable heuristics to evaluate an interactive system.

Keywords: usability, heuristics, iTV, Open-HEREDEUX.

1 Introduction

Interactive systems are growing in popularity and with the advent of technological
innovations, they occupies an important place in society' priorities. Thus, the improvement of interactive systems is a constant challenge. In the context of audience,
interactive systems are aimed to a broader spectrum of users each time. Such users
are, day by day, less "experts" in the application area who puts User eXperience (UX)
as a fundamental aspect for the success of systems software [1]. UX deals with all
facts, internal as well as external facts of the user and interactive systems, which
causes any feeling in the user who uses the interactive system in a specific context of
use. UX encloses different facets related to software quality such as: usability, accessibility, multiculturalism, among others. Bearing in mind that usability is one of the
most traditional UX facets, usability evaluation is one of the most important stages
inside the User-Centered Design [2]. Among the methods to evaluate the usability of
a software system, heuristic evaluation as an inspection method is the one of the most
efficient and most used methodologies. It is easy to implement, cheaper than others
(mainly those which require end users) and able to find many usability problems [3].
However, it may miss specific domain problems. That is why the use of appropriate
heuristics is highly significant.

C. Collazos, A. Liborio, and C. Rusu (Eds.): CLIHC 2013, LNCS 8278, pp. 55–58, 2013.

Taking into account the procedure for performing a heuristic evaluation, the activity related for defining the most appropriate set of heuristics and specific system features to be evaluated, is a key activity among those that make up the planning stage of the evaluation. Open-HEREDEUX (Open HEuristic REsource for Designing and Evaluating User eXperience) [4] has been created in order to cover all needs to analyze (based on a set of heuristics) UX and automate every possible part of the process, including the task of identifying the most appropriate heuristics for various systems.

This paper describes the process undertaken to match a set of heuristics for interactive digital television (iTV) applications into Open-HEREDEUX resource, specifically in *Open Repository* [4]. The main function of *Open Repository* is to store all the information necessary to achieve the widest source of heuristics, in order to select the most suitable set depending on specific aspects of the interactive system (functionalities, components, features), UX facets and/or the attributes of the standard ISO/IEC 25010 [5]. The following section describes the processes involved to enter a set of iTV heuristics to Open Repository are presented. Section 3 details the preliminary results. Finally, conclusions are presented.

2   Including the iTV Heuristics into Open-HEREDEUX

The process to enter the heuristics in Open Repository includes 6 tasks:

1. **Writing heuristics as declarative and interrogative sentences.** The heuristics to be included in Open Repository should be written in two modes: as declarative sentences and as questions. Initially, in [6] had been proposed a set of heuristics for iTV applications, however, some of them were not written in an declarative or interrogative mode. The task of writing the heuristics in declarative mode (for designers) and interrogative mode (for evaluators) was carried out manually.

2. **Discard duplicate heuristics.** New heuristics should be reviewed to avoid replications. Obviously, there are common iTV heuristics to other interactive software systems, so it was necessary to manually decide which heuristics for iTV applications were already registered in Open Repository, and then delete them. Open Repository stores 363 heuristics so, the set of iTV heuristics should be compared with all of them. This implies a big effort to include foreign heuristics.

3. **Associating features to the heuristics.** When designing an iTV application should be considered the physical features of interaction because users have an optimal vision at a particular distance from the screen; therefore, iTV applications should take into account screen resolution and contrast. Traditionally, users watch TV in an environment that is oriented to relaxation and comfort. However, nowadays users can access this medium in a wide variety of environments, from multiple devices and using different technologies. Open Repository currently presents the following features: having software part, having hardware part and simulate human behavior.

4. **Relationship of software and hardware components.** Open-HEREDEUX considers a hierarchy of components defined in a previous work. Some components were added considering the physical features of the interaction in the iTV

environment. The new hardware components created for covering iTV applications are: TV set, Set Top Box and remote control. The hierarchy currently stored in Repository consists of 57 components divided into software and hardware.

5. **Relationship of UX facets.** In [4] the authors conducted a study to determine the UX facets which are implicitly considered in the ISO/IEC 25010 [5]. Whereupon, Open Repository is based on the following UX facets: accessibility, reliability, desirable, emotion, findable, playability, plasticity, usability, useful, multiculturalism and communicability. The iTV heuristics are primarily related to the facets: usability, accessibility and findable.

6. **Evaluation of the degree of relevance of the heuristics.** UX degree is convenient to define different degrees of relevance to the heuristics that can be used in a specific interactive system. Thus the following degrees are defined: (1) **U degree**: the heuristics of the U degree are essential to assure that the user who will use the interactive system will get a positive experience. (2) **UU degree**: the heuristics of the UU degree are necessary to assure that the user who will use the interactive system will get a positive experience. (3) **UUU degree**: it is advisable to consider the heuristics of the UUU degree to assure that the user who will use the interactive system will get a positive experience. So, to define the degree of relevance of the iTV heuristics was made a consensus among seven (7) experts with the following profile: at least 3 years of experience in heuristic evaluations, knowledge about the areas of Human-Computer Interaction and User-Centered Design and basic knowledge about iTV applications. Each expert decided individually what UX degree best fits for each heuristic. Then, a consensus about the UX degree of each heuristic was obtained regarding the selection of all experts.

3   Preliminary Results

Due to space limitations, it is impossible to present the complete set of iTV heuristics and its specific UX degree. You can see the details in: (http://www.grihotools.udl.cat/openheredeux/cake_1_3/uxdegrees/showuxdegreeitv). A set of 142 heuristics have been entered in Open-HEREDEUX, which contribute to the design and evaluation of iTV applications. The designers (of user interfaces) of iTV applications can use Open Repository to obtain a set of useful design guidelines for functionalities, features, components or specific UX facets. These guidelines will be useful during the design process of the interaction of the user interfaces because these have been proposed to maximize UX. Evaluators of iTV applications can use specific heuristics related to functionality, feature, component, and facet to see if the interface being evaluated meets them or not. Likewise, Open-HEREDEUX (like heuristic evaluation) can be used at any stage of development of different interactive systems. The main benefit of Open-HEREDEUX is that it offers to UX/HCI practitioners a wide set of heuristics (from reliable sources and analyzed thoroughly), so that they can access quickly and easily the best set of heuristics. It is also important to note that Repository is open, so it will be constantly evolving.

4   Conclusions

Organizations that develop iTV applications, and other interactive software systems, would benefit from the valuable information of Open Repository, which are the heuristics. Currently Open Repository has 505 heuristics that can be applied to web systems, desktop applications, virtual assistants, public kiosk and iTV applications. By other hand, entering new heuristics to the resource is a task that demands a significant amount of time. The later considering the complexity of ensuring the proper and consistent introduction of the data. The main future activities are related to conduct case studies that demonstrate the correct recommendation heuristics by the tool. Also, it is necessary to conduct a case study with iTV applications, in order to verify that the recommended heuristics for this type of interactive system are appropriate.

Acknowledgments. This Project has been partially supported by the projects UsabiliTV: *"Framework para la evaluación de la usabilidad de aplicaciones en entornos de Televisión Digital Interactiva"* and the "Programa Nacional para Estudios de Doctorado en Colombia Año 2011, COLCIENCIAS".

References

1. Sharp, H., Rogers, Y., Preece, J.: Interaction Design Beyond Human - Computer Interaction, 2nd edn. Wiley, John & Sons, Incorporated (2007)
2. Otaiza, R., Rusu, C., Roncagliolo, S.: Evaluating the usability of transactional Web Sites. Presented at the Third International Conference on Advances in Computer-Human Interactions (ACHI 2010), Saint Maarten (2010)
3. Nielsen, J., Molich, R.: Heuristic evaluation of user interfaces. In: Proceedings of the SIGCHI Conference on Human Factors in Computing Systems: Empowering People, pp. 249–256 (1990)
4. Masip, L., Oliva, M., Granollers, T.: OPENHEREDEUX: open heuristic resource for designing and evaluating user experience. In: Campos, P., Graham, N., Jorge, J., Nunes, N., Palanque, P., Winckler, M. (eds.) INTERACT 2011, Part IV. LNCS, vol. 6949, pp. 418–421. Springer, Heidelberg (2011)
5. ISO, International Software Quality Standard, ISO/IEC 25010. In: Systems and Software Engineering - Systems and Software Quality Requirements and Evaluation (SQuaRE) - Systems and Software Quality Models (2011)
6. Solano, A., Rusu, C., Collazos, C.A., Arciniegas, J.: Evaluando aplicaciones de televisión digital interactiva a través de heurísticas de usabilidad. Ingeniare. Revista Chilena de Ingeniería 21, 16–29 (2013)

A Study about the Usability Evaluation of Social Systems from Messages in Natural Language

Marilia S. Mendes[1], Elizabeth Sucupira Furtado[2], Fábio Theophilo[2], Miguel Franklin[1]

[1] Federal University of Ceará (UFC), Fortaleza, CE, Brazil
[2] University of Fortaleza (Unifor), Fortaleza, CE, Brazil
{mariliamendes,elizabethsfur,
fabiotheophilo,miguel.franklin}@gmail.com

Abstract. Social Systems are dynamic systems, with features like interactivity, collaboration, sharing, diversity and a large number of users, various forms of access, focusing on human relationships and their emotions. In HCI (Human-Computer Interaction) there are several techniques that assess the usability of systems. However, such techniques do not consider the data collected from messages when users are interacting and expressing their feelings related to some difficulty in interaction. This paper presents a study about the usability evaluation of Social Systems from messages in Natural Language.

Keywords: Human Computer Interaction, Usability, Natural Processing Language, Social Systems.

1   Introduction

Consider Social Systems (SS) as the interactive environment in which people communicate, interact, collaborate and share ideas and information [1]. Also consider that users, when interacting with the system, praise, ask questions or complain about the system itself. Now imagine the amount of valuable data on the usability of the system that has been wasted by lack of analysis of users' posts? This paper focuses on the study about the evaluation of the usability in these kinds of systems. The motivation for this study can be summarized as follows:

(1) SS and its characteristics: Frequent message exchange. According to [2], recent studies have found that people are using messages in SS, especially to ask their friends some questions; Spontaneity. In SS, users who use the system continuously spontaneously comment about their use during interactions. According to [3], ask users what they think about a product, how they accomplish what they want, if they enjoy it, if the product is aesthetically appealing and if they face with problems while using it – it is an obvious way to get feedback. However, a question about the use of the system does not come "on time". Imagine the following scenario: suddenly a user is faced with a problem and he/she decides to comment on it in the system itself. He/she will exchange ideas with other users or even express his/her dissatisfaction about the problem; Expressing feelings. The author [4] claims that people turn to

C. Collazos, A. Liborio, and C. Rusu (Eds.): CLIHC 2013, LNCS 8278, pp. 59–62, 2013.
© Springer International Publishing Switzerland 2013

social media to express their feelings about important events in their lives, such as birthdays, etc. In [5], the author says that one way to deal with frustration induced by the computer is cashing it on the device itself or on other users expressing their feeling. He also states that when users are annoyed or irritated by something new, they overreact by typing things that they do not even dream of saying in person.

(2) Evaluation of usability: Many works have applied usability evaluation techniques to test these types of systems. In their evaluations, we did not find works that consider the content that users post to communicate, focusing exclusively on the actions of users, on their mistakes and opinions, usually collected from questionnaires or interviews; According to [1], evaluating the usability of SS communities requires different approaches from those applied to evaluate software used by only one user and, once molded, are relatively stable. Evaluating SS requires different approaches, as they have peculiar characteristics as said previously.

Thus, this paper makes an investigation about the usability evaluation in SS considering the study of Natural Language Processing (NLP). NLP aims to understand the language of human beings to create programs capable of interpreting messages encoded in natural languages [6].

2   Works Analyzed

Considering the context of the study on SS, we performed a search for papers about evaluation of usability in SS and about NPL. Usability is a measure that ensures that the products are easy to use, efficient, and enjoyable from the perspective of the user [5]. According to [5], evaluating usability in systems involves assessing the effect of the interface with the user by checking if he/she is satisfied or not to use the system. Satisfaction refers to the user's perceptions, feelings and opinions, usually obtained through written or oral questions [7]. In studies in which the context was the usability evaluation, it was noted that the techniques used were: questionnaire [8-11], interview [8, 10], heuristic evaluation [9, 10], Semiotics Inspection Method (MIS) [9, 11], Brokered Semiotic Inspection Method (MISI) [11], Communicability Evaluation Method (MAC) [12], usability testing [9], Inspection [9], Simplified Assessment of Accessibility (ASA) [9] and Method for Evaluation of Collaborative Systems (MASC) [13].

This study is based on the evaluation of the usability carried out by means of direct opinions of users. According to [11, 1, 5], the main techniques used to direct users to collect opinions and perceptions are questionnaires and interviews. The questionnaires are designed to collect data quickly (mostly quantitative) from many users. However, if the questions are not well made or are biased, they may induce or confuse the user. Furthermore, there is still the risk of the questionnaire being very long and users answering the questions without reading it carefully. Interviews, both individual and group (Focus Group, Brainstoming) are easy to understand, inexpensive and they have a more qualitative character. However, there is the effort to convening users. Also as in the questionnaires, the questions should be well formulated to achieve the desired goal and not to confuse the user. Another way to collect the user's opinion about the system is to apply the observation technique. This technique allows us to understand the user, his/her environment and how his/her tasks are carried out using a system [14]. Such technique, if it is not associated with interviews or questionnaires, does not allow the direct feedback from the user. There is so much effort into preparing visits, conducting and analyzing the results, as there is the risk of the user feels embarrassed to be seen. The embarrassment of the user is inherent in a usability test, inasmuch as it implies the observation of a person working with an interactive system [15]. The analyst must look for techniques and methods that reduce the level of embarrassment, ensuring the validity of the results.

Besides these reasons, the use of the techniques described here will not allow spontaneity to the user when he/she is using the system. The users talk about the system in use [2], however, it is difficult to assess the usability using the content posted, for the following reasons: the non-deterministic nature of the user would lead the appraiser not to know, among the posts, which one(s) refer(s) to usability; when the user referred to it during the interaction. In this scenario, a question can emerge: how representative is a user's complaint? This, of course, would lead to a great deal of effort from the appraiser, given the amount of users and contents posted to be investigated. One way to solve this problem would be by automating the process of selecting the contents posted. But then other questions arise: what types of contents posted by users are related to the system usability: doubts? complaints? "cheers"? How can NLP techniques assist in the extraction and processing of this content?

User satisfaction is bound to a field called Sentiment Analysis in NLP. According to [16] The Sentiment Analysis, also called Opinion Mining, is the field of study that examines people's opinions, feelings, evaluations, attitudes and emotions related to products, services, organizations, people, issues, events, etc. In the following paragraphs, we present some works that have used the NLP in order to perform a sentiment analysis. Such works were divided into three main categories:

(1) Text analysis for psychological classification: the work of [17] developed a text analysis that counts and categorizes words in the text. After analyzing texts of 400,000 words, including college students' surveys, messages from lovers and transcripts of the press, the authors concluded that the function of the words used have a strong relationship with the psychological state of people. One of the conclusions, for example, is that people who use a larger amount of personal pronouns in the first person, tend to be more depressed than those that do not use them.

(2) A study on population: The authors of [18] calculate the weight of each word related to mood (positive or negative) and represent the map by means of colors. With this, they analyze, for example, the effect of a speech delivered by the U.S. President, in real time, on the listeners connected to Twitter.

(3) Analysis of products and services: The work [19] have focused on the analysis of products and services by means of user feedback. In recent years, the use of sites for evaluation of products and services has become increasingly common. Sites (Booking, Decolar) provide a space for clients to disclose their personal opinions of products and services. The information on product reviews is of great interest both for businesses and consumers. Consumers are interested in the opinions of other consumers when they buy a product. These studies assess the context of the sentence and rank the views of users as positive or negative. In a hotel review site, for example, customers will describe their impressions of the hotel after their stay. In this analysis, expressions with the word "sheets" associated with the adjective "dirty" are classified as a negative impression about the hotel service. These works make their assessments

using words related to the service offered: hotels, for example, use sheets, room service and adjectives like dirty, clean, beautiful. There were no studies that use NLP to make an assessment of the SS usability.

3   Conclusion

This paper presented a study in progress of works about usability evaluation of SS and NPL. We believe that the messages of users provides a good insight into the use of the system. However, a tool for automatic analysis of posts would be as a *support* for evaluators to find what the users think about the system.

References

1. Pereira, R., Baranauskas, M., Silva, S.: Softwares sociais: umavisão orientada a valores. In: IHC 2010 (2010)
2. Sharoda, A., Lichan, H., Chi, E.H.: What is a question? crowdsourcing tweet categorization. In: CHI 2011 (2011)
3. Zhao, X., Sosik, V.S., Cosley, D.: It's complicated: how romantic partners use facebook. In: CHI 2012 (2012)
4. Brubaker, J., Swaine, F., Taber, L., Hayes, G.: The language of bereavement and distress in social media. In: AAAI 2011 (2011)
5. Rogers, Y., Sharp, H., Preece, J.: Interaction Design: Beyond Human Computer Interaction (2002)
6. Dias da Silva, B., Montilha, G., Rino, L., Specia, L., Nunes, M., Oliveira Jr., O., Martins, R., Pardo, T.: Introdução ao Processamento das Línguas Naturais (2007)
7. Jeffrey, R., Chisnell, D.: Handbook of usability testing: how to plan, design, and conduct effective tests. In: How to Plan, Design, and Conduct Effective Tests (2008)
8. Oliveira, A., Tavares, A., Oliveira, D., Pinheiro, M., Almeida, R., Darin, T.: Exposição de imagem no facebook - um estudo sobre a privacidade na rede social (2012)
9. Rodrigues, K., Canal, M., Xavier, R., Alencar, T., Neris, V.: Avaliando aspectos de privacidade no facebook pelas lentes de usabilidade, acessibilidade e fatores emocionais (2012)
10. Souza, L.G., Sippert, T.A.S., Cardoso, A.S., Boscarioli, C.: Análise da percepção e interação de usuários sobre privacidade e segurança no facebook. In: IHC 2012 (2012)
11. Terto, A., Alves, C., Rocha, J., Prates, R.: Imagem e privacidade: Contradições no facebook (2012)
12. Carvalho, J., Lammel, F., Dias da Silva, J., Chipeaux, L., Silveira, M.: Inspeçãosemiótica e avaliação de comunicabilidade: Identificando falhas de comunicabilidade sobre as configurações de privacidade do facebook. In: IHC 2012 (2012)
13. Madeira, K., Militão, J., Nóbrega, L., Santiago, L., Cursino, D., Matos, I.: Uma Avaliação do Orkut utilizando Personas sob a ótica da Nova Usabilidade. In: IHC 2008 (2008)
14. Millen, D.: Rapid Ethnography: Time Deepening Strategies for HCI Field Research (2000)
15. Cybis, W., Betiol, A., Faust, R.: Ergonomia e usabilidade: conhecimentos, métodos e aplicações, 352 p. Novatec, São Paulo (2007)
16. Liu, S.B.: Sentiment Analysis and Opinion Mining. Morgan & Claypool (2012)
17. Chung, C., Pennebaker, J.: The psychological functions of function words (2007)
18. Bollena, J., Maoa, H., Zengb, X.: Twitter mood predicts the stock market (2011)
19. Taboada, M., Brooke, J., Tofiloski, M., Voll, K., Stede, M.: Lexicon-based methods for sentiment analysis (2011)

A Quality Model for Human-Computer Interaction Evaluation in Ubiquitous Systems*

Rainara M. Santos[1],[**], Káthia M. de Oliveira[2], Rossana M.C. Andrade[1],[***],
Ismayle S. Santos[1],[†], and Edmilson R. Lima[1],[‡]

[1] Group of Computer Networks, Software Engineering and Systems (GREat)
Department of Computer Science, Federal University of Ceará. Fortaleza, Brazil
{rainarasantos,rossana,ismaylesantos,edmilsonrocha}@great.ufc.br
[2] University of Valenciennes, LAMIH, CNRS UMR 8201 Valenciennes, France
kathia.oliveira@univ-valenciennes.fr

Abstract. The improvement in computational device miniaturization
and in wireless communication has moved forward relevant advances in
ubiquitous systems development. Such systems are capable of monitoring environments and users in order to provide services as naturally as
possible. These systems offer new types of interactions, such as more
implicit and transparent exchanges with users. Thus, the ubiquitous systems present new challenges in quality evaluation of human-computer
interaction, as any assessment of quality should take into account the
peculiarities of these new types of interactions. This paper proposes
a quality model composed of specific characteristics and measures to
human-computer interaction quality evaluation in ubiquitous systems.
It also reports results obtained from a case study conducted to evaluate
an application based on this model.

Keywords: Ubiquitous Systems, HCI Evaluation, Quality Model.

1 Introduction

Ubiquitous Computing is a new computing paradigm that proposes the adoption
of computational devices in various sizes, shapes and functions to support users
daily activities. These systems will be everywhere around users, connected to
each other, and providing services which have to be as natural as possible. To
achieve this, the applications are embedded in everyday objects and capable of
monitoring user behaviour and environment [1]. To that end, the interaction
between the user and the system is of utmost importance and the quality of this
interaction has a direct impact on the use and adoption of the system. In this

---

* This work is a result of Maximum project supported by FUNCAP and CNRS under
grant number INC-0064-00012.01.00/12.
** Master Scholarship (MDCC/DC/UFC) sponsored by CAPES.
*** Researcher scholarship - DT Level 2, sponsored by CNPq.
† PhD Scholarship (MDCC/DC/UFC) sponsored by CAPES.
‡ Undergraduate Scholarship sponsored by PIBIC/CNPq.

C. Collazos, A. Liborio, and C. Rusu (Eds.): CLIHC 2013, LNCS 8278, pp. 63–70, 2013.

scenario, it is necessary to assure that these systems support user activities in a transparent way with little or no need for attention or input from a user.

Considering then the software quality evaluation, it is usually supported by a quality model that defines a set of characteristics (usually known as abilities of a system, such as usability and maintainability). Such characteristics are often organized in a hierarchical tree that starts with a generic definition and develops into measures that allow for the product assessment. The most commonly used quality model is the ISO 9126 Standard [2]. This standard specifies both the usability characteristic and measures for evaluating the Human-Computer Interaction (HCI). However, the nature of ubiquitous systems suggests that new quality characteristics should be taken into account. For example, an evaluation of ubiquitous systems should value an implicit and transparent user interaction over an interaction that requires direct input from the user.

This paper proposes a quality model to support HCI evaluation in ubiquitous systems. This model consists of characteristics and sub-characteristics that have impacts on user interaction quality, and measures capable of evaluating them for a particular system. It is important to mention that is not our goal to propose a complete model with all possible characteristics and their sub-characteristics, but rather to define a model with primary characteristics necessary for evaluating the HCI in ubiquitous systems. We called this model TRUU Quality Model, which means Trustability, Resource-limitedness, Usability and Ubiquity. We also present results obtained from a case study using this quality model.

2   The Quality Evaluation of HCI in Ubiquitous Systems

Mark Weiser's vision of ubiquitous computing is well expressed in his famous quote: "The most profound technologies are those that disappear. They weave themselves into the fabric of everyday life until they are indistinguishable from it"[3]. This paradigm includes services and information provision from a variety of computers that support users in everyday tasks. This support should be executed without users needing to be aware that they are interacting with various computer technologies. To achieve this goal, ubiquitous systems have to comply with challenging requirements such as autonomy, heterogeneity, coordination of activities, mobility and context-awareness [1].

To support these new systems, it is evident that the quality evaluation for ubiquitous systems should take into account new characteristics to evaluate the interaction between the user and the system. With the goal of identifying those characteristics, we performed a large literature review using a systematic mapping (SM) study. SM is an empirical methodology that provides a wide overview of a research area to establish if research evidence exists on a topic [6]. Using this methodology[1], we found 369 papers pertinent to the subject. By reading all abstracts, we selected 60 papers. With a deep reading of all papers, we selected only 18 of these papers ([4][5][9][10][11][12][13][14][15][16][17][18] [19][20][21][22][23]

---

[1] Details of this SM is out of scope of this paper, but it can be found in
   http://www.great.ufc.br/maximum/images/arquivos/protocolo.pdf

[24]) that discussed some quality characteristics specific to the HCI in ubiquitous systems. Then, we found that: *a*) Only two papers propose a model [5][4], but they do not present all important ubiquitous features (for example, [5] does not consider availability and [4] does not consider transparency); *b*) Some of them [17] [16], not organized in a model structure, focus in only specific aspects ignoring important aspects from ubiquitous systems (for example, calmness [21] and transparency [5]); and *c*) The papers present incomplete measures to assess the characteristics, with no formulas, interpretation or collection methods.

As a result of this analysis, we concluded they could be complementary, because some of them have important characteristic that the others do not have. Thus, it is important to organize a more complete quality model focused in all aspects mentioned by those authors and that should be considered in an evaluation of the HCI in ubiquitous systems.

3   The TRUU Quality Model

To define a quality model to evaluate HCI in ubiquitous systems, we decide to analyze and synthesize the 18 papers found in the literature (see in the previous section) concerning the subject. Based on these findings, we joined together the existing propositions in a single quality model, specific for HCI evaluation and called TRUU as presented in Table 1.

Table 1. The TRUU Quality Model

Characteristics	Sub-characteristics	References
Trustability	Security	[4] [9] [10] [11] [12] [13]
	Privacy	[10] [13] [5] [14] [15]
	Control	[13] [5]
	Awareness	[5]
Resource-limitedness	Device Capability	[12] [5] [16]
	Network Capability	[16]
Usability	Satisfaction	[11] [5] [16] [17] [18] [19]
	Ease of Use	[4] [9] [13] [17] [20] [21]
	Efficiency	[11] [5] [13] [17]
	Effectiveness	[5] [13]
	Familiarity	[5]
Ubiquity	Context-Awareness	[4] [11] [14] [9] [21] [22] [23]
	Transparency	[5] [23] [18]
	Availability	[4] [9] [13] [23]
	Focus	[13] [12] [5]
	Calmness	[13] [18] [21]

As shown in Table 1, we found characteristics that can evaluate any type of system (for example, satisfaction) and others specific to ubiquitous systems evaluation (for example, transparency). Then, we propose to group these specifics characteristics (called sub-characteristics in the model) into the TRUU characteristics. For space reasons, we have only detailed here the sub-characteristics from Ubiquity (that has aspects related to ubiquitous systems), as follows:

- Context-Awareness is the ability of the system to use context to provide relevant services to user, where relevancy depends on the user's task [7];
- Transparency is "the extension of the system which consists of hidden components in the physical space and interaction is performed through natural interfaces" [5];
- Availability is the system's capability to provide continuous access to information resources anywhere and anytime [23];
- Focus is the system's capability to maintain the user's focus on the main task [5]; and
- Calmness prevents humans from feeling overwhelmed by information [18].

To complete the TRUU Quality Model, measures were defined for evaluating each sub-characteristic. In this paper, we focus on the measures for context-awareness that were used in the case study presented in the next section. To define the measures we used the Goal-Question-Metric method [8] and considered suggestions presented in [23], [11], [4] and [25]. Using GQM, we define our goal as to "**analyse** ubiquitous systems **for the purpose of** evaluating quality **with respect to** context-awareness **from the point of view of** the user.

To define the questions and measures, we had to analyse the meaning of being context-aware. [7] defines context-awareness when a system uses context information to provide relevant services to the user. This context can include any information used to characterize the situation of an entity, which is a person, place, or object considered relevant to the interaction, including the user and applications themselves. Some aspects of context-awareness directly impact HCI quality and, therefore, they need to be evaluated. One aspect is the adaptation correctness, which means the system adapts in a correct way, providing services and information correctly. Some factors that can influence this correctness are: the context correctness, as if the context is wrong, the adaptation will be likely be wrong too; and the context changing frequency, as if the changes occur quite often, the adaptation may not take place before another change occurs [25].

Another aspect is the time taken to adapt, since the information and/or services should be delivered in a reasonable time to the user. Based on this analysis, we defined the questions and measures presented in Table 2. We note that to calculate the measure *Adaptation Correctness* and *Context Correctness*, we should identify which adaptations the ubiquitous system proposes to do (for example, adaptation for different devices) and also which context information they use for these adaptations (for example, the screen resolution context information to adapt the application behaviour in different devices). The resulting measures will be the average from the individual adaptation and context correctness. The interpretation values are an initial proposition, based on our own experiences and [25]. They will be refined after concluding more case studies.

Table 2. Questions and Measures

Questions	Measures		
	Name	**Formula**	**Interpretation**
What is the adaptation correctness degree?	Adaptation Correctness	$\frac{\sum_{i=1}^{N}(Ai/Bi)*100}{N}$ N=Number of adaptations Ai=Number of correctly performed adaptations i Bi=Number of performed adaptations i	Low - 0 to 25% Medium - 26 to 80% High - 81 to 100%
	Context Correctness	$\frac{\sum_{j=1}^{N}(Aj/Bj)*100}{N}$ M=Number of different context information Aj=Number of corrects collected context information j Bj=Number of collected context information j	Low - 0 to 25% Medium - 26 to 80% High - 81 to 100%
	Context Frequency	F=Frequency of changing	Low-minutes Medium-seconds High-milliseconds
What is the adaptation average time?	Adaptation Time	T=The time taken to adapt	Short - milliseconds Medium - seconds High - minutes

4 Case Study

The case study was performed using a mobile and context-aware application (MCAA) called GREat Tour [26]. This application is a tour guide for a large laboratory from Federal University of Ceará. This application runs on the visitor's mobile device and provides information about the laboratory's rooms that s-he is visiting, using texts, images and videos.

To collect the measures presented in Table 2, we first identified the context information and adaptations considered by GREat Tour. It presents two adaptations (N=2) and two context informations (M=2). The first adaptation is the laboratory map view according to user's location. To this purpose, the system identifies the room through the user's mobile device that reads the QR Code installed in all the doors of the laboratory's rooms. With this input, GREat Tour updates the user's map. The second adaptation is about showing media according to the device battery level. When the battery level is low (0-9%), only text appears, when it is medium (10-20%), texts and images are displayed and when it is high (21-100%), text, images and videos are displayed.

Thus, the *Adaptation Correctness* measure takes into account the laboratory map view (i=1) and the media view (i=2) as adaptations. The *Context Correctness* measure takes into account user's location through QR Codes (j =1) and battery level (j =2) as context information. The *Context Frequency* takes into account changes in location (F) and the *Adaptation Time*, the time required to show the new map to the user (T).

The data needed to compute the measures were collected both automatically and manually. Automatic data were recorded in logs that contain: the URL of map presented according to QR Code captured, the time taken to show the map after the capture of the QR Code, the hour, minute, second and millisecond that the context was collected to calculate the context changing, the device battery level and the media presented with this value. Manual data was collected in forms filled in by evaluators that followed users during the tours, observing if the application worked correctly, in other words, identifying if the system performed a correct adaptation with correct information.

Twelve users participated in the evaluation. All of them have experience with MCAA and are from computer science domain. Their tasks were divided into three laboratory tours, with each tour consisted of three rooms to be visited. The visit consisted of updating the user map and viewing all the available informations. Each tour was done with a device in different battery charge levels for the user to experience the batteries level-based adaptation. The twelve users were equally divided in three groups to execute the test with different sequences of battery level, as presented in Table 3.

Table 3. The performed scenarios in our evaluation

		Group 1	Group 2	Group 3
Tour	**Visited Rooms**	**Battery Level**		
1	Seminars Room Library Administrative Room 1	High	Low	Medium
2	Prototyping Room Software R&D Lab 1 Meeting Room	Medium	High	Low
3	Kitchen Administrative Room 2 Research Lab	Low	Medium	High

The final result was calculated by the average of all users tours. The result is shown in Table 4. The measures about correctness had high results, only one of them was not 100% (*Adaptation Correctness*). This happened because with adaptation i=1, the wrong map was displayed. We investigated this result and identified that the instability of the wireless network was a possible cause for this adaptation problem. The *Context Frequency* measure was low, because the result was in minutes, i.e, it takes about one minute for a context change to happen. If changes occur frequently, the adaptation may not take place before another change occurs, influencing measures of adaptation correctness. The *Adaptation Time* was short (milliseconds). It is interesting to note that this result was inferior than *Context Frequency*, favouring adaptation correctness.

Table 4. Results

Measures	Results		Interpretation
AdaptationCorrectness	when i=1, 96% when i=2, 100%	98%	High Correctness
ContextCorrectness	when j=1, 100% when j=2, 100%	100%	High Correctness
ContextFrequency	00:01:37		Low Frequency
AdaptationTime	539 ms		Short Time

Based on the collected results, we can see that the high degree of correctness and low adaptation time provide to the GREat Tour application, a good HCI, regarding context-awareness measures defined by the TRUU Quality Model.

5   Conclusion and Future Work

This paper presented a model for the HCI quality evaluation in ubiquitous systems. We also presented a case study focused on the context-awareness characteristic. Currently, we are working on the execution of several case studies in order to evaluate the whole model and also on qualitative evaluations to know the user's perception about the characteristics from our model.

Acknowledgments. We thank for the technical support to this work: CTQS - Software Quality and Testing Cell of the GREat.

References

1. Rocha, L.S., Filho, J.B.F., Lima, F.F.P., Maia, M.E.F., Viana, W., Castro, M.F., Andrade, R.M.C.: Past, Present, and Future Perspectives on Ubiquitous Software Engineering. Journal of Systems and Software (2012)
2. ISO/IEC 9126. Software engineering Product quality Part 1 (2001)
3. Weiser, M.: The Computer for the 21st Century. Scientific American (1991)
4. Lee, J., Song, J., Kim, H., Choi, J., Yun, M.H.: A User-Centered Approach for Ubiquitous Service Evaluation: An Evaluation Metrics Focused on Human-System Interaction Capability. In: Lee, S., Choo, H., Ha, S., Shin, I.C., et al. (eds.) APCHI 2008. LNCS, vol. 5068, pp. 21–29. Springer, Heidelberg (2008)
5. Scholtz, J., Consolvo, S.: Toward a Framework for Evaluating Ubiquitous Computing Applications. IEEE Pervasive Computing (2004)
6. Kitchenham, K., Charters, S.: Guidelines for Performing Systematic Literature Reviews in Software Engineering. EBSE Technical Report (2007)
7. Abowd, G.D., Dey, A.K., et al.: Towards a Better Understanding of Context and Context-Awareness. In: Gellersen, H.-W. (ed.) HUC 1999. LNCS, vol. 1707, pp. 304–307. Springer, Heidelberg (1999)
8. Basili, V., Rombacj, H.: Goal Question Metric Paradigm. Encyclopedia of Software Engineering 2 (1994)
9. Lee, J., Yun, M.H.: Usability Assessment for Ubiquitous Services: Quantification of the Interactivity in Inter-personal Services. In: International Conference on Management of Innovation and Technology (2012)

10. Abi-Char, P.E., Mhamed, A., El-Hassan, B., Mokhtari, M.: A Flexible Privacy and Trust Based Context-aware Secure Framework. In: Lee, Y., Bien, Z.Z., Mokhtari, M., Kim, J.T., Park, M., Kim, J., Lee, H., Khalil, I. (eds.) ICOST 2010. LNCS, vol. 6159, pp. 17–23. Springer, Heidelberg (2010)
11. Ranganathan, A., Al-Muhtadi, J., et al.: Towards a Pervasive Computing Benchmark. In: International Conference on Pervasive Computing and Communications Workshops (2005)
12. Wu, C.-L., Fu, L.-C.: Design and Realization of a Framework for HumanSystem Interaction in Smart Homes. Transactions on Systems, Man, and Cybernetics: Systems and Humans (2012)
13. Kemp, E.A., Thompson, A.-J., Johnson, R.S.: Interface Evaluation for Invisibility and Ubiquity: An Example from E-learning. In: International Conference on Human-Computer Interaction: Design Centered HCI (2008)
14. Jia, L., Collins, M., Nixon, P.: Evaluating Trust-Based Access Control for Social Interaction. In: International Conference on Mobile Ubiquitous Computing, Systems, Services, and Technologies (2009)
15. Sun, T., Denko, M.K.: Performance Evaluation of Trust Management in Pervasive Computing. In: International Conference on Advanced Information Networking and Applications (2008)
16. Zhang, Y., Zhang, S., Tong, H.: Adaptive Service Delivery for Mobile Users in Ubiquitous Computing Environments. In: Ma, J., Jin, H., Yang, L.T., Tsai, J.J.-P. (eds.) UIC 2006. LNCS, vol. 4159, pp. 209–218. Springer, Heidelberg (2006)
17. Cappiello, I., Puglia, S., Vitaletti, A.: Design and Initial Evaluation of a Ubiquitous Touch-Based Remote Grocery Shopping Process. In: International Workshop on Near Field Communication (2009)
18. Iqbal, R., Sturm, J., et al.: User-centred Design and Evaluation of Ubiquitous Services. In: International Conference on Design of Communication - Documenting and Designing for Pervasive Information (2005)
19. Ross, T., Burnett, G.: Evaluating the Human-Machine Interface to Vehicle Navigation Systems as An Example of Ubiquitous Computing. International Journal of Human Computer Studies (2001)
20. Weihong-Guo, A., Blythe, et al.: Using Immersive Video to Evaluate Future Traveller Information Systems. IET Intelligent Transport Systems (2008)
21. Wagner, S., Toftegaard, T., Bertelsen, O.: Requirements for an Evaluation Infrastructure for Reliable Pervasive Healthcare Research. In: International Conference on Pervasive Computing Technologies for Healthcare (2012)
22. Kim, H.J., Choi, J.K., Ji, Y.: Usability Evaluation Framework for Ubiquitous Computing Device. In: International Conference on Convergence and Hybrid Information Technology (2008)
23. Kourouthanassis, P.E., Giaglis, G.M., Karaiskos, D.C.: Delineating the Degree of Pervasiveness in Pervasive Information Systems: An assessment framework and design implications. In: Pan-Hellenic Conference on Informatics (2008)
24. Sousa, B., Pentikousis, K., Curado, M.: UEF: Ubiquity Evaluation Framework. In: Masip-Bruin, X., Verchere, D., Tsaoussidis, V., Yannuzzi, M. (eds.) WWIC 2011. LNCS, vol. 6649, pp. 92–103. Springer, Heidelberg (2011)
25. Cheng, B.H.C., et al.: Software Engineering for Self-Adaptive Systems: A Research Roadmap. In: Cheng, B.H.C., de Lemos, R., Giese, H., Inverardi, P., Magee, J. (eds.) Software Engineering for Self-Adaptive Systems. LNCS, vol. 5525, pp. 1–26. Springer, Heidelberg (2009)
26. Marinho, F.G., Andrade, R.M.C., Viana, W., Maia, M.E.F., Rocha, L.S., et al.: MobiLine: A Nested Software Product Line for the Domain of Mobile and Context-aware Applications. Science of Computer Programming (2012)

Tablet Use Patterns and Drivers of User Satisfaction: A Gender Approach

Marta Calderón and Gabriela Marín

School of Computer Science and Informatics, University of Costa Rica
{marta.calderon,gabriela.marin}@ecci.ucr.ac.cr

Abstract. Understanding what for, when and where Computer Science students at the University of Costa Rica use tablets and identifying gender differences in their use were our goals. An online survey, which included closed and open questions, was conducted. Results show that women use their tablet more for leisure and appreciate it for its usability. Non-working men use tablets also for leisure but value their functionality. Finally, working men are more interested in tablets as support to their work, and value usability more than non-working men.

Keywords: tablet use, gender, Costa Rica.

1 Introduction

Tablets, as well as smartphones, are becoming more popular in Costa Rica. Currently telecommunications operators offer packages including a tablet at relatively low prices. This change in the market place has motivated a higher rate of tablet purchases. According to [1], by November, 2011, 3.7% of the Costa Rica Internet traffic total share corresponded to non-computer devices, and 27.1% of the non-computer device traffic share was generated from tablets. By November 2012, Google presented an Internet traffic study which reveals that 30% of searches generated in Costa Rica using its web searcher came from mobile devices, particularly smartphones and tablets [2]. This fact represents an increase of 142% in only one year.

Around the world, people have adopted tablets as a new mean of interaction, using them for searching in the Internet, interacting with other people, downloading applications, watching videos, and playing, among others. Currently market research is the source of the most frequently published information about tablet ownership and use. Pew Internet & American Life Project has reported on percentage of adults owning tablets in United States (US) since summer of 2010. By September, 2010, 4% of US adults owned a tablet, but by August 2012 this percentage had grown to 25% [3]. In Spain, the Interactive Advertising Bureau (IAB) reported that in 2011 only 8% of the Spanish population owned a tablet, but it reached 23% by 2012, which made tablets the device with the highest rate-of-ownership growth in that country [4]. The 2012 IDG Global Solutions (IGS) Mobile Survey mentions that Information Technology (IT) professionals are especially prone to purchase and use mobile devices and, as

C. Collazos, A. Liborio, and C. Rusu (Eds.): CLIHC 2013, LNCS 8278, pp. 71–78, 2013.

with other technologies in the past, are key decision-makers when selecting mobile devices, operating systems, and applications [5].

Little have been published on what for, when and where are tablets used. Even more, gender differences have not been considered. Given this lack of information, we defined as our goal to create a picture of how Computer Science students at the University of Costa Rica use tablets today. We chose this target group because they are or will soon become decision-makers. We also want to identify whether there are differences between men and women use habits.

The structure of this paper is as follows Section 2 describes related work. Section 3 presents the methodology followed. Section 4 shows results obtained: general usage pattern results, usage patterns by gender, and usage satisfaction by gender. Finally, Section 5 describes the conclusions of this research study.

2   Related Work

According to [4], in Spain by September 2012, the most common tasks done using a tablet were using e-mail, accessing social networks and reading news on line, most of the times at home. From March to May 2012, IDG Global Solutions (IGS) conducted a global survey on mobile Internet use, in which more than 21,000 persons around the world participated [5]. According to data gathered, 92% of respondents are using tablets for web browsing, 81% for reading emails, and 77% for using tablet applications. These activities require a screen larger than the one smartphones have. They conclude that tablets became personal devices mostly used for personal activities [5].

In the academic field, Müller *et al.* [6] show the results of a research project in which 33 people participated. The three most frequently activities reported by participants are checking emails, playing games, and using social networks. Predominantly, tablets are used for personal activities related to fun and relaxation, which is consistent with the fact that users are within their home (couch and bed) when using tablets.

Previous research shows that there are differences among men and women when interacting with computers for solving problems or following instructions. [7-8]. None of the identified studies on tablet use consider differences between female and male users. What for, when and where each gender use them are open questions and their answers will very likely reflect each gender´s interests and concerns.

3   Methodology

In order to gather information about how students use tablets, we designed an on line questionnaire using Lime Survey. This instrument contained 18 questions related to personal use, work related use, software tools used, place of use, use time, and degree of satisfaction. The survey included closed and open questions, and was validated by three professors who are tablet users. Participants were recruited through e-mail and the Computer Science School Facebook page. In total, 44 students volunteered to participate. The criteria for participating were: being a Computer Science student at the University of Costa Rica (undergraduate or postgraduate) and owning a tablet.

When the survey was online, the 44 volunteers were invited to respond it. The survey resulted in 42 complete responses, which means a response rate over 95%.

Of all 42 participants who took part in the survey, 78.57% (33) were men and 21.43% (9) were women, which is representative of the distribution of female and male students studying Computer Science at our University. Their age range is from 16 to 51. While 61.9% (26) of participants study and work, 38.1% (16) only study. Half of the participants have used a tablet for one year or less and the other half from 13 months up to 3 years. Their average daily use time is 4.15 hours.

In order to find explanations to some of facts found through the survey, we additionally did follow up interviews to five students (3 women and 2 men).

4   Results

Survey results are presented in Section 4.1 to depict usage patterns for all the respondents. In Section 4.2 and 4.3 results are analyzed by gender.

4.1   General Usage Pattern Results

Figure 1 presents the different activities reported as performed using tablets. Results are organized in descending order by frequency of use.

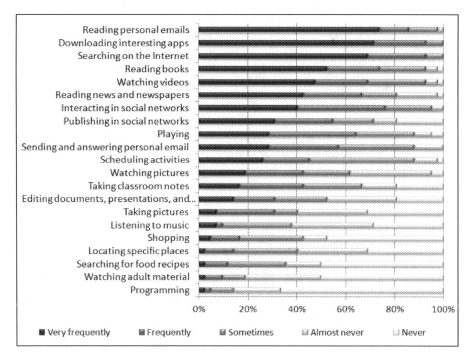

Fig. 1. Tablet personal usage by frequency of activity

The most popular activities are at the top (*reading personal emails, downloading interesting applications*, and *searching on the Internet*). These are the same top three most popular activities reported by IDG Global Solutions (IGS) [5]. Activities such as *locating specific places (GPS), searching for food recipes, watching adult material*, and *computer programming* seem to be done very infrequently by the surveyed student population.

Not all students work during their studies (38.1% do not work). From the ones that are currently employed (61.9%), most students report using their tablets at their workplace. *Sending and answering emails*, and *reviewing work documents* are the most popular work-related activities done with tablets.

Another interesting feature of tablet usage is the location or place where tablets are used. They are mainly used in relaxed comfortable locations (*bed* or *sofa*) but rarely used in the outdoors (*swimming pool, beach, at the sun* or *in the car*) except in *buses*.

Results obtained are similar to the ones reported by previous research [6]. This is in itself an interesting result since it highlights that tablet use patterns can be regarded as similar amongst individuals of different countries.

In Sections 4.2 and 4.3 we will analyze whether there are gender differences on tablet pattern usage.

4.2   Usage Patterns by Gender

Figure 2 presents the average frequency of the activities reported as performed using tablets, by gender. The average frequency was calculated using the following scale *Very Frequently* =4, *Frequently* = 3, *Sometimes* = 2, *Almost Never* =1, and *Never* =0. Thus, higher average frequency of use is reflected by greater distance from the center.

Some of the activities reflect similar behavior for both men and women, for example, *reading personal emails*, and *searching on the Internet*. Others exhibit behaviors that are foreseeable (at least in Latin American countries), like men using them more frequently to *watch adult material* than women, or women *searching for food recipes* more frequently than men. Some differences are important to take into account, like men more frequently *reading news and newspapers, reading books, scheduling activities, editing documents*, and *taking classroom notes* than women. This seems to suggest that men use their tablets to review content, and in more formal contexts than women.

Results that are surprising are that women utilize tablets more frequently than men for *playing*, and *listening to music*, and men use them more than women for *interacting in social networks*. Follow up interviews gave some light on these differences. Women argue that they like simple games, like cards and children games which can be easily downloaded as tablet applications, whereas men prefer more sophisticated games, and better sound quality music, deterring them from using their tablets for these activities.

Men´s more frequent interaction through social networks using their tablets can be related to the fact that they use tablets to review content. This was confirmed in follow up interviews.

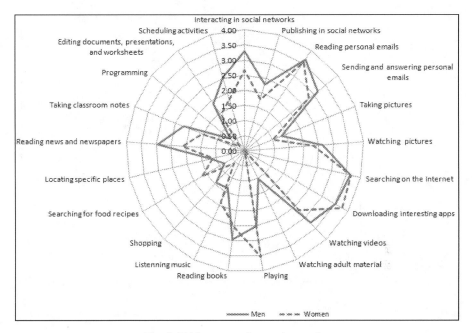

Fig. 2. Tablet personal usage by gender

Similar to results reported by Múller *et al.* [6], *bed* and *sofa* are the most popular places to use tablets. Figure 3 confirmed our belief that women use their tablet more often for relax activities than men. Moreover, men, more than women, use their tablets in more structured settings, like *table*, desk, *classroom*, and *office*.

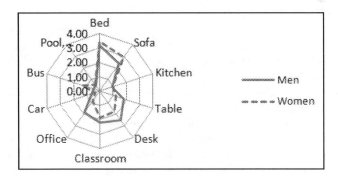

Fig. 3. Tablet usage location by gender

Women tend also to use tablets more at night than during the day than men (see Figure 4) and their average daily use time is slightly, but not significantly, higher (4.33 hours for women against 4.1 hours for men).

As tablet usage patterns are important for their design, drivers of user satisfaction may be valuable sources of information for human computer interaction specialists.

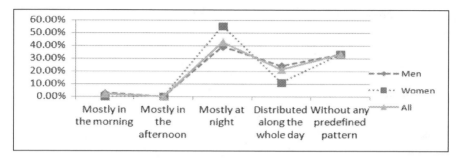

Fig. 4. Tablet daily use patterns by gender

4.3   Usage Satisfaction by Gender

Figure 5 shows that tablet owners are mostly satisfied or very satisfied with their tablet. However, satisfaction is slightly higher for men than for women.

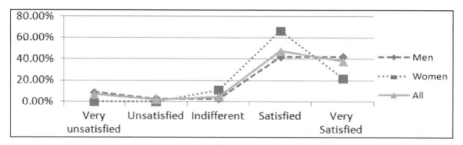

Fig. 5. Tablet user satisfaction by gender

What do you like most of your table? was an open question in our survey instrument. The answers were aggregated into tablet characteristics and the results were tabulated on Table 1, by gender and working status. Data on Table 1 were analyzed using correspondence analysis. Figure 6 shows a perceptual map reflecting graphically the differences among four user groups (represented in red). In a perceptual map, issues grouped together are said to be related. Therefore, *long* distance between user groups represents differences. Statistically, the resulting graph explains 88.69% of the data variation. From it, it can be seen that working women and non-working women are more similar between themselves than working men and non-working men. Moreover, it can be argued, based on Fig. 6, that women and men have different reasons for liking tablets.

After analyzing where characteristics are located in the perceptual map, graph axes were given names: the vertical axis reflecting the dichotomy between **leisure** and **work**, and the horizontal axis between **functionality** and **usability**. Women use their tablet more for leisure and appreciate it for its usability. Non-working men use tablets also for leisure but appreciate their functionality. Finally, working men are more interested in tablet as support to their work, and value usability more than non-working men.

Table 1. Frequency of tablet characteristics, by gender and working status

Code	Characteristic	Working women	Non-working women	Working men	Non-working men
APA	Application availability	1	2	5	3
BAL	Battery life	1	0	1	2
BOA	Book availability/reading	0	0	3	0
EAI	Easy to access information	1	2	0	0
EMA	Email alarms	0	0	0	1
EUE	Enjoyable user experience	0	0	0	1
EUS	Easy to use	1	0	2	0
GAM	Games	0	0	0	1
GPS	Support to geog. location	0	0	0	1
IOD	Integration with other devices	0	0	2	0
MEE	Memory expandability	0	0	0	1
OSF	Operating System Features	1	2	1	0
PRF	Processor features	0	0	4	5
PRI	Price	0	0	2	2
SCI	Screen interface	1	1	1	1
SCR	Screen resolution	0	0	3	0
SCS	Screen size	0	0	1	0
SDF	Software development facilities	0	0	1	0
STW	Support to work	0	0	0	1
TAR	Tablet always ready	0	0	0	2
TRA	Transportability	0	1	6	4
TSI	Tablet size	2	1	6	1
TWE	Table weight	1	0	2	1
VER	Versatility	0	1	4	1

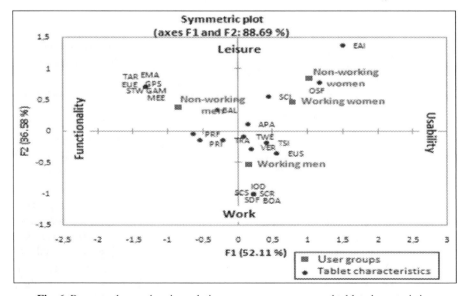

Fig. 6. Perceptual map showing relation among user groups and tablet characteristics

We asked students about what they do not like from their tablets. The three top characteristics they responded are related to *limitations imposed by the tablet operating system* (women 33% and men 21%), *user interface* (women 22% and men 21%), and *short battery life* (women 33% and men 12%). These statistics are congruent with the fact that men are slightly more satisfied with their tablets than women (Figure 5). Only men (6%) complained of poor processor performance and software development limitations, consistent with the fact that men use tablets as a support to work tasks.

5   Conclusions

Computer Science students using tablets follow the same usage patterns reported by previous market and academic research efforts. This in itself is interesting since it can contribute to consolidate results obtained elsewhere. However, our more important contribution is that different tablet usage patterns between men and women were depicted. Women use tablets mostly at home, in bed and sofas, and are more interested in usability and leisure. Men use them more than women at work and in more structured scenarios. They are more interested in functionality and tablets as support to work activities. These findings might be important for future tablet design and human computer interface customization.

References

1. comScore, Tablets Account for Nearly 40 Percent of Non-Computer Web Traffic in Brazil and Colombia, http://www.comscore.com/Insights/ Press_Releases/2011/12/Tablets_Account_for_Nearly_40_Percent_ of_Non-Computer_Web_Traffic_in_Brazil_and_Colombia
2. CRHoy, Estudio muestra qué buscan los ticos en Google y cómo lo hacen, http://www.crhoy.com/estudio-muestra-que-buscan-los-ticos- en-google-y-como-lo-hacen/
3. Pew Internet & American Life Project, 25% of American Adults Own Tablet Computers, http://www.pewinternet.org/~media//Files/Reports/2012/ PIP_TabletOwnership_August2012.pdf
4. Interactive Advertising Bureau, IV Estudio Anual IAB Spain Mobile Marketing, http://www.iabspain.net/wp-content/uploads/downloads/2012/09/ IV-Estudio-IAB-Spain-sobre-Mobile-Marketing-Versión- Completa1.pdf
5. IDG Global Solutions, How Mobility is Disrupting Technology and Information Consumption, http://idgknowledgehub.com/mobileidg/idg-mobile-survey/
6. Múller, H., Gove, J.L., Webb, J.S.: Understanding Tablet Use: A Multi-Method Exploration. In: Proceedings of the 14th International Conference on Human-Computer Interaction with Mobile Devices and Services Mobile, HCI 2012, pp. 1–10. ACM, New York (2012)
7. Beckwith, L., Burnett, M., Wiedenbeck, S., Grigoreanu, V.: Gender HCI: What About the Software. J. IEEE Computer 39(11), 97–101 (2006)
8. Jonsson, I.M., Harris, H., Nass, C.: How Accurate Must an In-Car Information System Be? Consequences of Accurate and Inaccurate Information in Cars. In: Proceedings of the SIGCHI Conference on Human Factors in Computer Systems, CHI 2008, pp. 1665–1674. ACM, New York (2008)

Studying the Relationships between the Management of Personal Data Privacy and User Interface

Sandra R. Murillo[1] and J. Alfredo Sánchez[2]

[1] Universidad Popular Autónoma del Estado de Puebla,
Puebla, Puebla 72410,
México
[2] Universidad de las Américas Puebla,
Cholula, Puebla 72810,
México
sandrarocio.murillo@upaep.mx
j.alfredo.sanchez@gmail.com

Abstract. Despite technological efforts that have been implemented so that users can navigate on the Internet with increased control of the privacy of their information, results are not always as expected. The lack of information security culture and the inconsistency of preventive or corrective interfaces among common applications create confusion in people and insecurity about the use of their information by third parties. Some researchers have suggested that the application of learning styles to build interfaces can facilitate users' cognitive ergonomics. This paper presents the current relationship between the management of personal data privacy, user interface and learning styles through an ethnographic study. These relationships suggest a model to navigate among these components and improve user experience.

Keywords: Human factors, usability, security.

1   Introduction

There are important factors to be considered by users to manage their privacy on the Internet. Most users usually move quickly through default settings when they install or use software without reading the details. Internet sites maintain records that indicate whether privacy notices were read and the terms and conditions of use of personal data were accepted. However, users in general are not aware of the consequences of the accepted terms. When any software has been used for a while, the risks associated with the information provided at the beginning or during usage are not evident. Users interact with applications on a daily basis without being certain of what can happen if a third party misuses the data they capture. Today, there is no universal format for presenting the elements of a website privacy notice and the associated risks for the owner of the information. There are international laws applied in some countries that require legal language to show how stored information can be

C. Collazos, A. Liborio, and C. Rusu (Eds.): CLIHC 2013, LNCS 8278, pp. 79–89, 2013.

handled by a third party. The interface is critical at all times to support user decisions on this issue. Some research papers have suggested the application of learning styles for building interfaces; however there is no evidence of their application in the management of the information privacy of Internet users. This paper reports initial findings on how learning styles and user interfaces are related with the way the privacy of personal information is managed. The paper is organized as follows: Section 2 discusses the main issues in the realm of information privacy on the Internet. The main problem is defined in Section 3, whereas Section 4 describes the methodology followed in our study. Our initial results are presented in Section 5 and discussed further in Section 6. Finally, Section 7 presents the conclusions derived from our study.

2   Internet Privacy

The increasing use of diverse technologies has made it possible for personal user information to be used for purposes other than those originally intended. Sometimes personal data may be transferred to other entities without   the owner's explicit consent. From the perspective of personal data protection, the owner of the data has the right and the freedom to decide what to communicate, when and to whom, maintaining control over their personal information at all times [1]. Due to problems that have arisen (identity theft, fraud, etc.), international organizations have proposed mechanisms for software developers to include options that strengthen user privacy. On the technical side it has been possible to detect and provide solutions [2]. However, the low level of information security culture and unsuitable interface designs cause the average user to remain unaware of the solutions available [3].

Some systems offer privacy options in legal language that is unclear to the average user. Some people, by curiosity or necessity, take time to understand the concepts to strengthen measures to prevent or solve privacy problems. However, each type of software handles terms and symbols differently, thus causing frustration and rejection. A paper published in 2011 [4] suggested that making the risks that can affect the privacy of users of a Web application explicit, can help them make better decisions regarding this issue. However, for some users, static design discourages the use of these applications. Considering human factors in safety mechanisms has increased the efficiency of the proposed solutions.

During the last years, researchers have found some relationship between security topics and human factors as shown in Table 1.

From the perspective of neurophysiology and psychology, some articles have proposed cognitive, affective and physiological models with features that indicate how people perceive their interactions and respond to learning environments [9]. These are referred to as learning styles and have contributed to the educational process in recent years [10]. Some research projects have suggested their application to generate interfaces that facilitate users' cognitive ergonomics [11-13].

Table 1. Relationship between security topics and human factors

Research	Security factors	Human factors
An online study [5] showed that policy formats do have significant impact on users' ability to both quickly and accurately find information and on users' attitudes regarding the experience of using privacy policies. In another study, [6] participants did not continue reading the full policy when the information they sought was not available at short notice.	Policy formats	Unclear language, finding information, recognizing personal data.
However, users interviewed in Switzerland and India feel responsible for protecting their sensitive documents themselves, instead of relying on cloud storage providers. It shows that end-users have a strong belief, fueled by media stories and hacker stereotypes, that the Internet is intrinsically insecure [7].	Cloud storage	Unreliable
A job studied both the relationship between demographics and phishing susceptibility. Gender and age are two key demographics that predict phishing susceptibility [8].	Phishing susceptibility	Demographics
A study revealed than people really get angry about apps in social networks because they take and redistribute their personal information without consulting them first , in spite of the fact that it is specified somewhere in the social network [4].	Third-party access and redistributing data	Concern, anger, frustration

3   The Problem

The number of Internet users is increasing globally and this trend is expected to continue. In a national study with adults, it is observed that 13% said that information owners are responsible for the proper treatment of personal data. However, 31% could not define what personal data may be and only 19% always read the privacy notice of a web site, whereas 62% of those surveyed take over 4 minutes to read it. 61% do not know how their information will be handled in the social networks they are involved in [14].

These data reveal the following:

1. The growing number of Internet users lack information security culture in the area of data privacy.
2. Users have interfaces at their disposal to manage their privacy but do not use them.
3. Adult Internet users do not know how to apply security measures.

The legal terms in extensive privacy policies generate feelings of insecurity, uncertainty or indifference in users as to the processing of personal data. People do not know the real effect of what they share on line. Sometimes, they delegate responsibility for information management to the default options presented by the interface. So far, learning styles have not been considered to define the elements of privacy notices and their relationship with users. This also limits developers to creating interfaces that meet the goal of helping surfers in their decision making.

Multidisciplinary teams work to create interfaces and follow the legal requirements in each country. Software components and graphic elements have been associated with the concept of security. However, it cannot be ignored that there are different user profiles and static designs cannot cover all of them. Learning styles and design theories have been applied to create virtual environments to improve cognitive ergonomics. Color has been used as a quantifiable variable of human-computer interaction. So far, no other variables have been proposed to associate privacy with personality features.

3.1   Objective

The goal of this research is to identify the current relationship among the factors of the privacy management of personal data, user interface and learning styles.

3.2   Specific Objectives

1. To identify the current relationship between technical and human factors in Internet use and information privacy
2. To establish a relationship between privacy management and learning styles of web users.
3. To recognize the factors considered by users to be critical in managing their privacy.

4   Methodology

A quantitative analysis of data was performed. It was based on a parametric analysis using frequencies and correlation coefficients [13]. Frequency analysis and the Pearson correlation method were applied using the SPSS software to identify relationships between the study variables. A survey was applied to 280 students in a University in the city of Puebla, México, with a total population of 8157 students. Their participation was voluntary and compensation was not offered to participants. For a significant sample, the number of respondents was proportional to the total number of students for each major discipline as seen in Table 2. Only 275 surveys were used for the study. This presents a confidence level of 95% with an error of 5.81%.

4.1 Demographics

Demographic factors considered are: Major, semester, sex and age, as seen in Tables 2, 3, 4 and 5. The study inquired about the predominant learning style of the participants [15]. It recognized the significant presence of the four styles in the entire sample as shown in Table 6.

4.2 Study Questions

The study was designed to include three blocks in a third version of the survey:

1) Demographics.
2) Learning style: People selected their significant learning style based on:

   Active: Improviser, discoverer, adventurous and spontaneous. I prefer to do and experiment.
   Pragmatic: Impatient, experimenter, practical, effective and realistic. I prefer to try things out to see their functionality.
   Reflective: Prudent, cautious, researcher, collector, analytical and thorough. I prefer to observe and reflect.
   Theoretical: Methodical, logical, objective, critical and structured. I prefer to understand the concepts, ideas and relationships between them.

3) We presented a seven-point Likert (0: not at all important, 6: extremely important). We asked eight questions to understand their knowledge of privacy and we asked four questions to better understand some interface factors:

   Rate the importance in degree of the following activities when you browse any website:

   1. Identifying who collects your personal data.
   2. Knowing what they are being used for.
   3. Applying basic privacy measures.
   4. Reading the privacy notice.

5. Recognizing the main elements in a privacy notice.
6. Knowing the federal law about protecting personal data in your country.
7. Identifying how you will be notified if there are changes in the privacy notice in a site.
8. Identifying options for checking, modifying or canceling your data.
9. Simple language.
10. Attractive interface design.
11. Customizing the browser.
12. Suggestions about interface customization according to your tastes.

The relationship among these variables was analyzed using SPSS [16]. The reliability of the instrument was validated based on Cronbach's Alpha of 0.863 as shown in Table 7 and Table 8.

4.3   Correlation between Study Variables

Based on the results obtained by the Pearson Correlation Coefficient, the most significant relationships of the factors that users consider important to manage personal data privacy when browsing any website can be summarized as shown in Fig. 1.

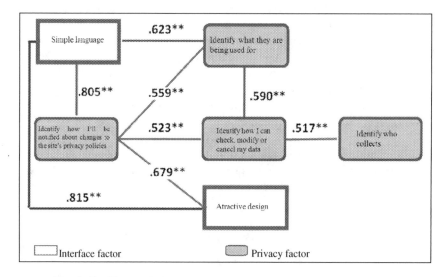

Fig. 1. Significant relationships based on Pearson Correlation Coefficient

At the end of the survey, participants were asked about the overall rate of satisfaction about the decisions made regarding privacy of data (0 is no satisfaction and 10 is complete satisfaction). The results are presented in Table 9.

Table 2. Major

Major	Frequency	Percentage
Accounting and Senior Management	6	2.3
Agronomic Engineering	3	1.1
Architecture	10	3.6
Automotive Design Engineering	10	3.6
Bionic Engineering	5	1.8
Biotechnology Engineering	2	0.7
Business Administration	5	1.8
Chemical Engineering	2	0.7
Cinema and Audiovisual Production	7	2.5
Civil Engineering	5	1.8
Communications	6	2.2
Computer Systems Engineering	6	2.2
Design and Advertising Production	13	4.7
Economics	1	0.4
Electronic Engineering	4	1.5
Environmental Engineering	11	4
Financial and Stock Market Management	1	0.4
Gastronomy	9	3.3
Humanities	2	0.7
Industrial Engineering	11	4
Intelligence Business	4	1.5
International Relations	3	1.1
International Trade	10	3.7
Journalism	1	0.4
Law	6	2.2
Marketing	5	1.8
Mechatronics Engineering	9	3.3
Medicine	71	25.8
Institutional Management	5	1.9
Nursing	7	2.5
Nutrition	4	1.5
Odontology	9	2.9
Pedagogy	2	0.7
Philosophy	3	1.2
Physiotherapy	5	1.8
Political Sciences	1	0.4
Psychology	5	1.8
Psychopedagogy	2	0.7
Veterinary Medicine and Zootechnics	4	1.5

Table 3. Semester

Semester	Frequency	Percentage
1	46	16.7
2	8	2.9
3	47	17.1
4	15	5.5
5	57	20.7
6	17	6.2
7	48	17.5
8	12	4.4
9	18	6.5
10	5	1.8
11 or more	2	0.8
Total	275	100

Table 4. Sex

Sex	Frequency	Percentage
Male	136	49.5
Female	138	50.5
Total	275	100

Table 5. Age

Age	Frequency	Percentage
17	4	1.5
18	26	9.5
19	50	18.2
20	63	22.9
21	56	20.4
22	35	12.7
23	29	10.5
24	7	2.5
25	5	1.8
Total	275	100

Table 6. Learning style

Learning style	Frequency	Percentage
Active	78	28.4
Reflexive	73	26.5
Theoretical	56	20.4
Pragmatic	68	24.7
Total	275	100

Table 7. Summary of case processing

		N	%
Cases	Excluded	275	100
	Valid[a]	0	0
	Total	275	100
a. Deletion based on all variables in the procedure.			

Table 8. Statistical reliability

Cronbach's Alpha	N elements
0.863	12

5   Results

5.1   General Results

These are the main results from this study:

- A majority of users, regardless of their learning style, present a similar handling of their privacy.
- 85.8% identify their personal information.
- 56.7% know what a Privacy Notice is.
- 87.6% were not aware of the existence of federal laws about the protection of personal data.
- 18.9% apply online measures to maintain the privacy of their personal data.
- A majority of users recognize that it is important to know what third parties do with their personal data; however, no security measures are applied (Fig. 2)
- 18.2% of participants read the online privacy notices of the pages they visit and take from one to four minutes. (Fig. 3)
- 48% customize their browser (Fig. 4 and Fig. 5)
- Most respondents have a high level of satisfaction about their privacy decisions. (Fig. 6)
- To manage their privacy, 81.5% would like their tastes for interface customization to be taken into consideration.
- 81.5% expect a simple format and attractive design in a privacy notice.

The four most significant correlations based on the Pearson correlation coefficient are:

1. Interfaces must have attractive and simple language (.815 **)
2. Changes to the privacy policies should be displayed in plain language (.805 **)
3. Changes to the privacy policies should be displayed in an attractive design (.679 **)
4. The use of personal data should be displayed in plain language (.623 **)

Table 9. Overall rate of satisfaction

Valids	Frequency	Percentage
1	4	1.5
2	31	11.3
3	21	7.6
4	25	9.1
5	41	14.9
6	34	12.4
7	45	16.4
8	51	18.5
9	22	8.0
10	1	0.4
Total	275	100.0

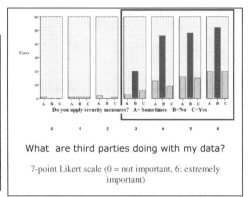

What are third parties doing with my data?

7-point Likert scale (0 = not important, 6: extremely important)

Fig. 2. Use of personal data and security measures

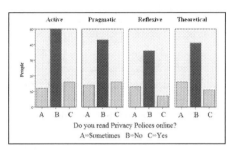

Fig. 3. Learning style and reading privacy notices

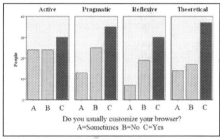

Fig. 4. Learning style and browser customization

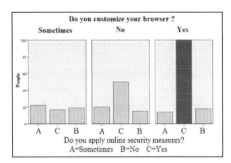

Fig. 5. Browser customization and security measures

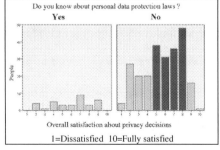

Fig. 6. Overall satisfaction about privacy decisions

6 Discussion

The study was conducted with students from different undergraduate semesters of all the existing majors in a university. These characteristics assume a particular approach and management of information technology and also assume a certain socioeconomic level for a geographic area. Future studies could include people who are outside the set explicitly considered in this ethnographic study and compare the results. The previous experience of the participants was taken into account and a usability study with any particular type of software was not applied. This study considered a particular learning styles theory because there are already tools that allow for their identification; however, it is recognized that there are other theories to consider.

In the context of information privacy, interface design is crucial for managing the security of Internet users (see Fig. 1). Current interfaces do not support the process of understanding the basics of this topic, as some authors have suggested, but the results are not as expected. The absolute values of Pearson's correlation coefficients may suggest a hierarchical navigation between these components. Most respondents customize their interface when surfing on the Internet; however, no security measures are applied (see Fig. 5). This presents areas of opportunity for the proposal of alternate mechanisms to present this information and encourage its application.

This study distinguishes similar behavior in how people manage issues about personal data regardless of their learning style (see Fig. 3). Users want customization preferences to be taken into account (see Fig. 5). What effect would be generated by considering these preferences in the management of personal data? Would different behavior be observed, in accordance with their learning styles?

This study recognizes the importance of privacy for users and the high degree of satisfaction they claim to have with their decisions regarding privacy (see Fig. 6). However, no security measures are applied by users (see Fig. 5), so this is an obvious problem. But users want their configuration tastes to be taken into account (see Fig. 5) and recognize the importance of protecting their data (see Fig. 2). This means that there are a number of opportunities to propose alternative mechanisms to provide information and promote the implementation of security measures.

7   Conclusions

This study shows that the factors related to user interface should be considered from the perspective of security. It would be desirable to provide implementation guidelines that relate the critical factors of information privacy and interface design topics. All learning styles show similar behavior in regards to the interfaces of current privacy management. There is an obvious problem in that users are satisfied with their decisions about the handling of their information without safety measures. Understanding how people perceive and respond to these interactions could bring benefits in terms of increasing their security and in allowing developers to have a framework to improve user experience and system usability.

References

1. Lindskog, H.: Web Site Privacy with P3P. Wiley (2003)
2. Cranor, L., Langheinrich, M., Marchiori, M.: The Platform for Privacy Preferences 1.0 (P3P1.0) Specification, http://www.w3.org/TR/P3P/

3. Garfinkel, S., Faith, L.: Security and Usability. Designing Secure Systems That People Can Use. O'Reilly (2005)
4. King, J., Lampine, A., Smolen, A.: Privacy: Is there an app for that? In: Symposium on Usable Privacy and Security, http://dl.acm.org/citation.cfm?id=2078843
5. Gage, P., Cesca, L., Bresee, J., Faith, L.: Standardizing Privacy Notices: An Online Study of the Nutrition Label Approach, http://www.cylab.cmu.edu/files/pdfs/tech_reports/CMUCyLab09014.pdf
6. McDonald, A., Reeder, R., Gage, P., Lorrie Faith, L.: A Comparative Study of Online Privacy Policies and Formats, http://www.springer.de/comp/lncs/index.html
7. Ion, I., Sachdeva, N., Kumaraguru, P., C'apkun, S.: Home is Safer than the Cloud! Privacy Concerns for Consumer Cloud Storage, http://cups.cs.cmu.edu/soups/2011/proceedings/a13_Sachdeva.pdf
8. Sheng, S., Holbrook, M., Kumaraguru, P., Faith, L.D.: Who Falls for Phish? A Demographic Analysis of Phishing Susceptibility and Effectiveness of Interventions, http://dl.acm.org/citation.cfm?id=1753383
9. Hannaford, C.: Cómo aprende tu cerebro. In: PAX, pp. 20–25 (2011)
10. Secretaría de Educación Pública. Manual de estilos de aprendizaje, http://www.dgb.sep.gob.mx/informacion_academica/actividadesparaescolares/multimedia/Manual.pdf
11. Jaime, M.-A., Gustavo, R.-G.: Patrones de Interacción. Una solución para el Diseño de Retroalimentación Visual de Sistemas Interactivos, http://ccc.inaoep.mx/~grodrig/Descargas/InteraPatternToCIC.pdf
12. Velasco-Santos, P., A. L.-C.-T.: Diseño de agentes pedagógicos a partir de los estilos de aprendizaje; una perspectiva a través del color. In: Memorias del IV Congreso Mundial de Estilos de Aprendizaje (2010) ISBN 978-607-7533-66-5
13. Velez, O., Solano, D., Zúñiga, L., Argüello, M., Aguado, J., Aldana, J., López, P., Alarcón, A.: Los estilos cognitivos y el aprendizaje maquinal en el diseño de interfaces inteligentes adaptativas. Universidad del Sinú. Centro de Investigaciones y Desarrollo, 3–6 (2005)
14. Asociación Mexicana de Internet AMIPCI. Estudio de Protección de Datos personales entre usuarios y empresas, http://www.amipci.org.mx/?P=editomultimediafile&Multimedia=95&Type=1
15. CHAEA. Cuestionario Honey Alonso de Estilos de aprendizaje, http://www.estilosdeaprendizaje.es
16. Malhortra, N.: Investigación de Mercados, 5th edn. Pearson Prentice Hall (2008)

HCI with Chocolate: Introducing HCI Concepts to Brazilian Girls in Elementary School

Cristiano Maciel[1], Sílvia Amélia Bim[2], and Clodis Boscarioli[3]

[1] Universidade Federal de Mato Grosso (UFMT) , Instituto de Computação (IC)
Laboratório de Ambientes Virtuais Interativos (LAVI), Cuiabá, MT 78050-970, Brazil
[2] Universidade Tecnológica Federal do Paraná (UTFPR), Departamento de Informática
(DAINF) Curitiba, PR, 80230-901, Brazil
[3] Universidade Estadual do Oeste do Paraná (UNIOESTE)
Colegiado de Ciência da Computação – campus de Cascavel, PR 85819-110, Brazil
cmaciel@ufmt.br, sabim@uftpr.edu.br,
clodis.boscarioli@unioeste.br

Abstract. In order to attract women to the area of computing there are several initiatives in the Brazilian context. The project named *Meninas Digitais* (Digital Girls) is one of them. In this paper we discuss one experiment carried out in the context of this project. A Computer Science Unplugged activity (*The Chocolate Factory*) was performed in a Brazilian state school. The activity was about HCI design and was done with nine teenagers. Most of these girls do not have a computer, so they rarely use one. The experiment was their first contact with a topic related to concepts of HCI design, and the girls succeeded in the activity giving interesting solutions for the problem situations presented. The experiment showed that it is possible to introduce some activities to elementary school students so as to present HCI and promote courses in the area of Computing.

Keywords: HCI design, Computer Science Unplugged, Women in IT.

1   Introduction

The participation of women, where computing is involved, both in academia and in the job market has been much discussed. The IEEE Women in Engineering (WIE) [6] is dedicated to promoting women scientists and engineers, facilitating the recruitment and retention of women in technical disciplines globally.

The Brazilian Computer Society (SBC) has held, in the last years, the WIT (Women in Information Technology) [7] - a workshop to discuss subjects related to gender and IT. The Project *Meninas Digitais* [5] was created from discussions in the WIT, the main objective of which is to promote Computing and technology for girls from ten to sixteen at the end of elementary school or middle school, in order to generate interest in the area and motivate them to choose Computer Science as a career. Several activities are being carried out by various collaborators throughout Brazil such as lectures, technical visits and other initiatives. This paper presents one of these

C. Collazos, A. Liborio, and C. Rusu (Eds.): CLIHC 2013, LNCS 8278, pp. 90–94, 2013.

activities and brings some of its results and insights from one pilot experiment conducted in a public school in the city of Cuiabá-MT/Brazil. The activity, based on one of the proposals of the Computer Science Unplugged Project for the area of Human-Computer Interaction [2], was held with nine teenager students. Among our aims we seek to promote strategies in the area of Human-Computer Interaction (HCI) for the dissemination of a computing career for girls.

2   HCI Design Without Computer

The Computer Science Unplugged Project [1] has as its main objective the promotion of Computer Science as an interesting, engaging and intellectually stimulating subject for young people. Therefore, they have developed a series of activities addressing various topics in Computing including that of HCI – Human-computer interaction. "The Chocolate Factory activity" addresses human interface design [2]. It was translated into Brazilian Portuguese to be used by the girls in our experiment [3].

The setting is that of a chocolate factory and the users are the Oompa-Loompas[1] who have a number of characteristics that need to be considered: they cannot write, they cannot read and have very bad memories. Because of this, they have difficulty in remembering what to do in order to run the chocolate factory and things often go wrong. The goal of the activity, which consists of five tasks[2] is to design a new factory that is supposed to be very easy for them to operate [2].

The experiment was conducted at the Municipal School of Basic Education Lenine Campos Póvoas on the outskirts of Cuiabá (MT) with nine girls who accepted the invitation and brought the consent form signed by a parent or legal guardian. Three affinity groups were formed with three girls in each. The tasks were explained with the use of slides and a data projector. At the end, a questionnaire was used to record the profile of the participants and the careers they intend to follow. The results and insights obtained are reported as follows.

3   Findings

3.1   Questionnaire Analysis

The participants were thirteen (three), fourteen (three) and fifteen (three) years old.
Only three out of nine girls have computers at home and consequently, the use of computers is very rare. Only two girls answered that they use computers daily whilst the others reported using one hour occasionally. Therefore, access to the internet is also very sporadic.

When they use the computer it is mainly for communication purposes. They also use it for studying activities, games, music and photos. The girls were questioned

---

[1] The scenario is based on British book Charlie and the Chocolate Factory, by Roald Dahl, published in 1964, which inspired a movie of the same name, released in 2005.

[2] But only four of them were conducted because of time limitations.

about their knowledge of Computer Science courses. Eight of them replied "None" and one answered "I know that it is a new course and has a lot of good things.". This reflects their experience with computers.

When questioned about their professional careers ("Do you already know the professional career would you like to follow? Why?") two girls gave generic answers, one of them answered "No" and the others mentioned Accounting, Engineering, Agronomy Architecture (three of them) and Law. Most of them justified their choices because they like Math.

From these answers it is possible to identify a potential interest for the Computer Science area since three of the girls like Math. We can also interpret that Human-Computer Interaction could be an area of interest since it can be related to Architecture in the sense that both are concerned with people and interaction and deal with creative activities [4]. Consequently, it shows that it is necessary to increase activities in Brazilian schools, especially public schools, to promote the Computer Science area.

3.2   The Experiment Analysis

Task 1 – The door design: All of the groups chose the labeled door. On the other hand, although they chose one of the types suggested by the experiment, the girls mentioned that an automatic door would suit better. In spite of having a very creative insight they didn´t register this on the worksheet showing that it is difficult to break paradigms. This result indicates that is necessary to explore with the girls that for each different situation it would be important to study possibilities but also to propose new ones.

Task 2 – Redesign the stove: Two groups used labels to indicate the controls of the stove, one with numbers and the other with letters. Although they are interesting mapping designs the suggestions are not suitable for the Oompa-Loompas since they can't read. Group three´s design was the best one offering a direct mapping without using labels. Their suggestion is the safest one because it prevents the Oompa-Loompas from burning their sleeves when reaching across the burners to adjust the controls.

Task 3 – Visual warning system: All the groups respected the Oompa-Loompas' traffic light color pattern (yellow means stop, red means go, and green means slow). However, each group arranged the commands in a different order. Interestingly, none of them used the visual layout of classic traffic lights. This can be interpreted as showing a good understanding of the differences between the girls' (designer) world and the Oompa-Loompas' (user) land.

Nevertheless, Group 1 put the colors in the reverse order: green, yellow and red. Group 2, in its turn, put the commands in the same order of classic traffic lights – stop, slow down and go. The solution proposed by Group 3 doesn't use either color or the command order of classic traffics light.

Task 4 – Cupboard design: Group 1 used a color pattern with written labels to identify each type of utensil. Probably, the color alternative is a consequence of the previous activity that dealt with colors. It is interesting the way that the girls applied the concept of reuse, very common in Computer Science activities, without knowing it. However, this is not a good solution, firstly because the Oompa-Loompas can't read, and secondly because they will need to memorize which color represents which utensil, and they are very bad at remembering things. This alerts us to the fact that reuse may not suit design problems especially when the users have specific characteristics. In their turns, groups 2 and 3 proposed better solutions using figures to represent the utensils.

In all of the activities the girls proposed solutions ignoring the Oompa-Loompas' reading disability.

4   Discussion and Future Works

The girls showed a high level of involvement in the activities and despite having little experience with computers they are curious about the computing area and accepted taking part in order to know more about it.

Their responses to the design problems proposed in the activities were interesting, especially in the first activity where they came up with a creative option by suggesting the automatic door. However, it is important to explore the importance of considering the users' characteristic in designing the solutions. One possible way is to use their interest in Architecture to construct the relationship between a building project and an interface project.

The kind of activity conducted in this experiment is very important to give an opportunity to disseminate the Computer Science area not only for the girls, but also for the teachers and the schools. New experiments are been planned in other Brazilian cities and schools in order to continue the work of bringing knowledge about computing and also to collect more evidence on the Brazilian reality in relation to the use and knowledge of technology.

References

1. CS-Unplugged Principles. Disponível em, http://csunplugged.org/unplugged-principles
2. CS-Unplugged. Human Interface Design - The Chocolate Factory Activity. Disponível em, http://csunplugged.org/sites/default/files/activity_pdfs_full/unplugged-19-human_interface_design_0.pdf
3. Maciel, C., Bim, S.A., Boscarioli, C.: A fantástica fábrica de chocolate: Levando o sabor de IHC para meninas do ensino fundamental. In: Proceedings of the 11th Brazilian Symposium on Human Factors in Computing Systems (IHC 2012), pp. 27–28. Brazilian Computer Society, Cuiabá (2012)
4. Rogers, Y., Sharp, H., Preece, J.: Interaction Design: Beyond Human Computer Interaction, 3rd edn. John Wiley & Sons (2011), http://www.id-book.com/

Borrowing a Virtual Rehabilitation Tool for the Physical Activation and Cognitive Stimulation of Elders

Alberto L. Morán[1], Felipe Orihuela-Espina[2], Victoria Meza-Kubo[1],
Ana I. Grimaldo[1], Cristina Ramírez-Fernández[1], Eloisa García-Canseco[1],
Juan Manuel Oropeza-Salas[2], and Luis Enrique Sucar[2]

[1] Facultad de Ciencias, UABC, Ensenada, B.C., México
[2] Instituto Nacional de Astrofísica, Óptica y Electrónica, Tontanzintla, Puebla, México
{alberto.moran,mmeza,ana.grimaldo,al302126,
eloisa.garcia}@uabc.edu.mx
{f.orihuela-espina,juan.oropeza,esucar}@inaoep.mx

Abstract. We explore the use of a virtual rehabilitation platform as the interaction means for physical activation and cognitive stimulation of elders. A usability evaluation of actual and projected use of the tool suggests that this could be feasible to perform. Elders perceived the use of the evaluated tool as useful (93.75/100), easy to use (93.75/100) and pleasurable to use (91.66/100) during an actual activation and stimulation session. Previous experience on the use of computers by the participants did not significantly impact on their usability perception for most of the included factors, with the sole exception being the perception of anxiety. This is an encouraging result to reuse and adapt technologies from "close" domains (e.g., virtual rehabilitation). In addition, this can reduce development times and cost, and facilitate knowledge transfer into the domain of physical activation and cognitive stimulation of elders.

Keywords: Usability study, virtual rehabilitation, cognitive stimulation, physical activation, elders.

1 Introduction

Life expectancy of human beings has increased over the past few decades. However, longer living impacts the quality of life at an old age, as an increased age is usually associated with a decline in the physical and cognitive abilities of elders. This decline may be due to, or aggravated by, the existence of wear or disease. Technology can potentially help alleviating the decline in innovative ways [e.g. 2, 9, 10].

Under this latter goal we are interested in investigating the development of technology that supports the physical activation of the elderly and their cognitive stimulation in order to meet both the physical and the cognitive aspects concurrently. In this paper, rather than developing new technology, we explore an alternative approach; reusing existing exogenous technology. We present an exercise that takes a virtual rehabilitation system of the upper limb, and evaluates its usability as an interaction means for the physical activation and cognitive stimulation of the elderly. If

C. Collazos, A. Liborio, and C. Rusu (Eds.): CLIHC 2013, LNCS 8278, pp. 95–102, 2013.
© Springer International Publishing Switzerland 2013

succeeding, that will open the door for cross-breeding technologies across domains. In particular, in this first study we are interested in usability; i) obtaining the elders' perception regarding the use of the rehabilitation tool as an interaction means for an activation and cognitive stimulation application, and ii) whether the novelty of the device and application has an effect on the perception of users that have (or have not) experience on the use of this or similar technology.

2   Background

Virtual Rehabilitation. Brain injury resulting from stroke or palsy often leaves patients with motor impairment. Several rehabilitation therapies are available for motor restoration [1]. Among them, virtual reality based therapies have recently become a valid alternative, an option that is now known as virtual rehabilitation [2]. Virtual rehabilitation commonly involves concealing rehabilitation exercises as serious games. By now, a few tens of virtual rehabilitation platforms have been developed [3]. For the experiments presented in this work we rely on Gesture Therapy (GT) [3, 4], a low cost virtual rehabilitation platform for the upper limb illustrated in Figure 1. GT has demonstrated clinical value similar to occupational therapy in two different trials but with an edge on motivation [4, 5].

Fig. 1. The Gesture therapy virtual rehabilitation platform for the upper limb. Gesture Therapy characteristic gripper permits controlling of the user avatar through tracking of the colorful ball as well as monitoring gripping forces by an incorporated pressure sensor. Serious games of the platform as the one pictured encourage repetitive exercises beneficial for motor rehabilitation.

Cognitive Stimulation for the Elderly. Some age-related diseases, such as Alzheimer, are accompanied by the patient's cognitive decline. Cognitive decline is characterized by difficulties in one of the following: memory and learning, attention and concentration, problem solving and abstraction, language comprehension and word finding and/or visual functioning [6]. Cognitive stimulation is a non-pharmacological intervention for people with dementia by which a range of enjoyable activities bring forth general stimulation for thinking, concentration and memory usually in a social setting, such as a small group [7]. Frequent participation in cognitive stimulation activities reduces the risk of suffering a cognitive decline related disease, improving the patient's cognitive functioning and behavior [8]. To deliver cognitive stimulation activities, a number of platforms use serious games as a mean [9, 10, 11], thus being an ideal environment for testing transference from virtual rehabilitation technology. We have chosen cognitive stimulation as a target application to integrate with physical activation (through virtual rehabilitation games) because the population affected by cognitive decline, i.e. elders, is also the population most affected by stroke.

3   Usability Evaluation

Participants were 32 elderly (age mean±std: 64.96±6.31 years) recruited from a local municipal third age support group in Ensenada, Mexico. They live an independent life and have no apparent cognitive problems. The cohort was exposed to a subset of three games of the GT platform (steak cooking, window cleaning and fly killer), and asked to evaluate them in terms of perceived usefulness, ease of use and user experience. Our main goal was to assess usability aspects of the recycled technology. In addition, their previous experience, or lack of it, with technologies was logged to use it as a factor for explaining the evaluation results. The hypothesis was that older adults with previous experience with technology were to perform better in the serious games, and consequently, likely evaluate the games more graciously.

After a brief explanation of the experiment objectives, participants signed a consent form. Upon commencing the experimental session they answered a questionnaire regarding their demographic data and their familiarity with technologies. A 2-minute demonstration of the use of the platform followed by a 3-minute free playing familiarization (training) was given to each participant. Afterwards, the subjects played the games for 15 minutes each. The order in which the games were played was randomized. User interaction with the games was monitored with software tool Camtasia (TechSmith, USA). Finally, participants were asked to fill a 29-element extended TAM-based questionnaire [12]. The questionnaire addresses 5 sections: perception of usefulness, perception of ease of use, intention of use (should the system be available), anxiety experience during system usage, and user experience. Most elements of this questionnaire (22 items) are 5-point Likert scales. The remaining 7 items have a 3-option answer where the preference of games has to be ranked.

4   Evaluation Results

4.1   Performance on the Game

The average scores per game and category (technology experienced and inexperienced) are presented in Table 1. As expected, subjects that reported having previous experience on computer or game console use slightly outperformed those reporting lacking experience. Independent-samples t-tests were conducted to compare the performance on each game of the two groups of subjects.

There were no significant differences in the scores for any of the performed games for experienced subjects and inexperienced subjects, conditions; steak cooking: $t_{30} = 0.52$, $p = 0.303$; window cleaning: $t_{30} = 0.9$, $p = 0.187$; and fly killer: $t_{30} = 1.07$, $p = 0.146$. This suggests that previous technological experience does not give an edge start when using GT that reflects significantly on performance. Thus, we consider that the proposed games and controller gripper allow experienced and inexperienced subjects to achieve similar performances while conducting this experiment.

Table 1. Score averages per game and previous experience with technology category

Experienced Subjects (ES) (N=17)				Inexperienced Subjects (NES) (N=15)			
	Steak cooking	Window cleaning	Fly killer		Steak cooking	Window cleaning	Fly killer
Avg.	141	789.35	58.64	Avg.	128	739.26	50.6
S.D.	76.89	142.86	20.86	S.D.	63.23	173.48	21.83

4.2   Overall Usability Perception

To evaluate the perception of usefulness, ease of use and user experience, we scored the questionnaire items of each category following an approach similar to that used in the System Usability Scale (SUS) [13]. This way, scores greater that 62.5 will indicate a trend towards users agreeing (75) or completely agreeing (100) that the proposed feature is present or supported by the application under evaluation.

Table 2. Evaluation results concerning the overall usability perception of subjects per groups using the 22 Likert scale items only

Overall Usability Perception *(alpha = 0.05, p = 0.065)*			
Experienced Subjects (ES)		**Inexperienced Subjects (NES)**	
Median	94.32	**Median**	88.63
IQR	6.81	**IQR**	21.59
Mean Rank	18.9	**Mean Rank**	13.8

Table 2 presents a summary of the results categorized according to participants' having or not experience using the computer. As hypothesized, technology experienced subjects granted higher notes (median = 94.32 (Interquartile Rank (IQR) = 6.81)) than inexperienced subjects (median = 88.63 (IQR = 21.59)); however, the effect was found to be marginally not significant at the 0.05 level (Mann-Whitney U: p=0.065). These results suggest that a) most subjects tend to agree (A) or completely agree (CA) that the application provides or promotes each of the evaluated factors (i.e. usefulness, ease of use, etc.), and b) that previous experience on computer or game console use does not have an effect on the subjects' overall usability perception of the platform.

Table 3. Summary of Mann-Whitney U Test results on the perception of usefulness, ease of use, intention of use, user experience and anxiety by category

Experienced Subjects (ES)		Inexperienced Subjects (NES)	
Perceived Usefulness *(alpha = 0.05, p = 0.153)*			
Median	93.75	Median	100
IQR	12.5	IQR	6.25
Mean Rank	14.9	Mean Rank	18.3
Perceived Ease of Use *(alpha = 0.05, p = 0. 254)*			
Median	93.75	Median	93.75
IQR	12.5	IQR	25
Mean Rank	17.6	Mean Rank	15.3
Perceived Intention of Use *(alpha = 0.05, p = 0. 173)*			
Median	100	Median	100
IQR	12.5	IQR	25
Mean Rank	18	Mean Rank	14.8
Perceived User Experience *(alpha = 0.05, p = 0. 440)*			
Median	91.66	Median	95.83
IQR	4.16	IQR	33.33
Mean Rank	16.8	Mean-Rank	16.2
Perceived Anxiety *(alpha = 0.05, p = 0. 017)*			
Median	100	Median	87.5
IQR	4.16	IQR	29.16
Mean Rank	19.8	Mean Rank	12.8

4.3   Perceived Usefulness, Ease of Use, Intention of Use and User Experience

To further scrutinize usability aspects we further analyzed perception of usefulness, ease of use, anxiety and user experience of elders (see Table 3). As can be seen in Table 3, both groups of participants perceived high values for the aspects considered in this study. On the one hand, contrary to hypothesized, the inexperienced group (NES) granted higher notes than the experienced group (ES) on the perceived usefulness (NES: Median = 100 (IQR = 6.25), ES: Median = 93.75 (IQR = 12.5)) and on the perceived user experience (NES: Median = 95.83 (IQR = 33.33), ES: Median = 91.66 (IQR = 4.16)). Furthermore, both groups granted very similar notes on the perceived ease of use (ES: Median = 93.75 (IQR = 12.5), NES: Median = 93.75 (IQR = 25)) and on the perceived intention of use (ES: Median = 100 (IQR = 12.5), NES: Median = 100 (IQR = 25)). However, subsequent Mann-Whitney U tests showed no significant differences at the 0.05 level for any of the four described aspects: perceived usefulness (p=0.153), perceived ease of use (p=0.254), perceived intention of use (p=0.173) and perceived user experience (p=0.440). These results suggest that experience on computer or game console use does not have an effect on any of these aspects regarding the use of the GT platform.

4.4   Perceived Anxiety

Subjects from both groups reported low levels of anxiety while using the GT platform as shown in Table 3. As hypothesized, experienced subjects granted higher notes on low levels of anxiety (Median = 100 (IQR = 4.16)) than inexperienced subjects (Median = 87.5 (IQR = 29.16)). The results of a Mann-Whitney U test suggest that the mean ranks differ in the same directions as the medians, and that the effect was found to be significant at the 0.05 level (p=0.017), which suggests that prior experience affects subjects' anxiety perception when first confronted with the GT platform.

4.5   Participants' Thoughts on the Provided Technology

Usefulness. Subjects were also asked to categorize the games regarding the usefulness they perceived after playing with them (items 5 and 6). Overall, fly killer was perceived as the most useful (41.74%), followed by steak cooking (31.11%) and window cleaning (27.15%). A possible explanation could be that elders from both groups perceived fly killer as the game that put more challenges on them (discussed in the next subsection). Interestingly, in private communications a rehabilitation consultant already told us he perceives the fly killer to be the most useful for rehabilitation.

Ease of Use. Subjects were also asked to categorize the games regarding the ease of use they perceived after playing with them (items 11 and 12). Overall, window cleaning was perceived as the most ease to play game (40.13%), followed by steak cooking (32.71%) and fly killer (27.16%). Not surprisingly in the light of the higher scores, despite the similar scoring system across games, however unexpected since window cleaning is arguably the hardest task being the only requiring bidirectional movements, a game designed for advanced stages of motor rehabilitation therapies. A possible explanation could be that elders from both groups perceived window cleaning as the game with the most familiar activity (e.g. Inexperienced Subject 10 (NES10) said that cleaning the window "is something that I usually do, but I am not used to kill mosquitoes that way"), and that they perceived fly killer as the most-challenging game (e.g. NES15 said "it is the hardest game to play, […] you are required to move the sprayer towards the mosquito and press ["the handle"] to actually spray it").

User Experience. When asked to categorize the games considering which was the most fun to play (item 29), overall, fly killer was perceived as the most fun to play game (41.50%), followed by steak cooking (31.43%) and window cleaning (27.07%). Perhaps, elders felt a sense of satisfaction after killing the mosquito (e.g. NES6, while playing, said "take that! you won't be able to bite me anymore!); that they did not perceived window cleaning as a fun activity (e.g. NES1 said "I like to cook, but I don't like to clean the windows!"); and that being fly killer perceived as the most challenging activity, participants were proud of their scores, and interested in knowing how they have done in comparison to the others (e.g. after finishing the game, subject ES3 said "72 mosquitoes, wow! … what is the highest score?").

Intention of Use. Regarding projected frequency of use and session duration (items 15-16) 29 participants (90.6%) said that they would use it at least twice or thrice a week (Figure 2.A), and 27 participants (84.3%) declared that they would use it for 10 minutes or more during each projected session (Figure 2.B).

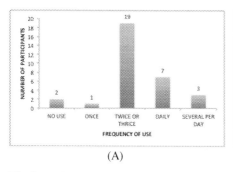 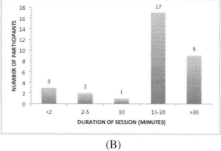

(A)   (B)

Fig. 2. Participants' intention of use regarding frequency of use (A) and duration of session (B)

Anxiety Levels. Anxiety levels could be inversely related to the subject's familiarity with controllers such as the proposed one (i.e. gripper), which in turn may influence the subject's expectations regarding the use of and interaction through the device. Although subjects from both groups found the system easy to use, experienced subjects were more at ease with the device (e.g. Experienced Subject 5 (ES5) said that "the gripper is similar to the Wii[mote] controller", and ES14 said that s/he has "already used this [a similar] controller with his/her grandson's [PlayStation] game console"), while some inexperienced subjects did not knew what to expect or what to do with the proposed device (e.g. NES15 said "In the beginning I did not know what this thing [referring to the gripper] was for or how to use it … that made me nervous, but once I understood how to use it, I felt much better", and NES8 said "I did not knew what to expect").

5   Discussion and Conclusions

Regarding the reutilization of a virtual rehabilitation tool as the interaction means for the physical activation and the cognitive stimulation of elders, our main findings are that i) subjects perceived the proposed games as useful, easy to use and capable of generating a pleasurable or fun user experience with a low anxiety level, and (ii) that subjects expressed their intention to use the games if available. This is an encouraging result to recycle technologies from "close" domains, which can reduce development times and cost, as well as facilitating knowledge transfer. Furthermore, we found that experience on the use of computer and game consoles does not impact the user's overall usability perception of the platform, nor on the specific factors; with the exception of the anxiety level.

Concerning the particular approach, we consider that i) GT activities foster the elderly upper limb physical activation and their cognitive stimulation (at an initial level); ii) the proposed application allows for assessing the elders' performance on the activity and provides feedback to render them aware of how they did in the exercise; iii) the elements (contents) of the activity are presented to them in an appropriate manner (metaphors) that eases their perception and learnability; and iv) the proposed device (gripper) provides the elderly with a simple but effective interaction channel.

These results provide promising evidence towards the feasibility of integrating virtual rehabilitation and cognitive stimulation technologies to concurrently support the physical (re) activation of the elderly and their cognitive stimulation. However, it is necessary to further identify specific elements from each domain and to determine how they relate among them in order to generate a more general design solution that aids in the development of useful, easy to use and pleasurable physical activation and cognitive stimulation applications. We aim to address this in our future work.

References

1. Langhorne, P., Coupar, F., Pollock, A.: Motor recovery after stroke: A systematic review. Lancet Neurology 8, 741–754 (2009)
2. Levin, M.F.: Can virtual reality offer enriched environments for rehabilitation? Expert Reviews of Neurotherapeutics 11(2), 153–155 (2011)
3. Sucar, L.E., Orihuela-Espina, F., et al.: Gesture therapy: An upper limb virtual reality-based motor rehabilitation platform. Submitted to IEEE Transactions on Neural Systems and Rehabilitation Engineering, 1–10 (2013)
4. Sucar, L.E., Velázquez, R.L., et al.: Gesture therapy: A vision-based system for upper extremity stroke rehabilitation. In: 32nd Annual International Conference of the IEEE Engineering in Medicine and Biology Society (EMBS), pp. 3690–3693. IEEE Press, Buenos Aires (2010)
5. Orihuela-Espina, F., Fernández, I., et al.: Neural reorganization accompanying upper limb motor rehabilitation from stroke with virtual reality-based gesture therapy. Topics Stroke Rehabilitation 20(3), 197–209 (2013)
6. Levy, R.: Aging-associated cognitive decline. International Psychogeriatrics 6(1), 63–68 (1994)
7. Woods, B., Aguirre, E., et al.: Cognitive stimulation to improve cognitive functioning in people with dementia. Cochrane Database of Systematic Reviews: Plain Language Summaries, page Abstract only (2012)
8. Meza-Kubo, V., Morán, A.L., Rodríguez, M.D.: Bridging the gap between illiterate older adults and cognitive stimulation technologies through pervasive computing. Universal Access in the Information Society, 1–13 (2012) (in press)
9. Tárraga, L., Boada, M., et al.: A randomised pilot study to assess the efficacy of an interactive, multimedia tool of cognitive stimulation in Alzheimer's disease. Journal of Neurology, Neurosurgery and Psychiatry 77(10), 1116–1121 (2006)
10. Gamberini, L., Alcaniz, M., et al.: Cognition, technology and games for the elderly: An introduction to elder games project. PsychNology Journal 4(3), 285–308 (2006)
11. Meza-Kubo, V., Morán, A.L.: UCSA: A design framework for usable cognitive systems for the worried-well. Personal and Ubiquitous Computing 17(6), 1135–1145 (2013)
12. Venkatesh, V., Davis, F.D.: A theoretical extension of the technology acceptance Model: four longitudinal field studies. Manage. Sci. 46(2), 186–204 (2000)
13. Brooke, J.: SUS: A "quick and dirty" usability scale. In: Usability Evaluation in Industry. Taylor and Francis (1996)

How Do You Understand Twitter?:
Analyzing Mental Models, Understanding and Learning about Complex Interactive Systems

Víctor M. González and Rodrigo Juárez

Department of Computer Science,
Instituto Tecnológico Autónomo de México, Mexico City, Mexico
{victor.gonzalez,miguel.juarez}@itam.mx

Abstract. The aim of this investigation is to identify and understand the relations between the people's mental models and their performance and usability perception about a complex interactive system (Twitter). Our study includes the participation of thirty college students where each of them was asked to perform a number of activities with Twitter, and to draw graphical representations of the mental model about it. The participants have either none or at least a year of expertise using Twitter. We identified three typical types of mental models used by participants to describe Twitter and found that the level of expertise had a major impact on performance rather than the mental model style defining the understanding about the system. Furthermore, and in contrast, we found that usability perception was affected by the level of expertise.

Keywords: mental models, HCI, Twitter, complex interactive systems.

1   Introduction

In recent years a massive increment has occurred in the design, production and usage of technology. This boom of technology creates a need of bearing in mind certain aspects that were not so important before to both, users and programmers. One of these aspects is the way in which people interact with the system, i.e. the Human-Computer Interaction (HCI).

At early stages of computing application development, most final users were the programmers themselves or people with a high technical knowledge. As years passed, the "final user" concept suffered diversification and today, the definition of user, implies a person with few or null conceptual knowledge about the system she uses.

This diversification process has profound implications for designers and programmers, because they have to consider how the system is understood, conceptualized and internalized in the mind of the users in order for it to be valued, used efficiently and achieving the purposes for which it was designed. Among the challenges, one should understand and consider that an interactive design has to depart from allowing users easily define a mental representation to communicate with programmers and designers' aims and ideas and vice versa (the Gulf of Execution and the Gulf of Evaluation suggested by Donald Norman).

C. Collazos, A. Liborio, and C. Rusu (Eds.): CLIHC 2013, LNCS 8278, pp. 103–110, 2013.

Cognitive sciences had shown that it is quite difficult to understand the world in a direct way, instead we use a representation of the world, elaborated in our minds, to act like a medium [1]. The most used representations among researchers are mental models [2] and those are the lenses that will use to frame our study.

1.1   Mental Models (MM)

Johnson-Laird defines mental models as a series of psychological representations of real, hypothetic or imaginary situations whose structure corresponds to the structure of the event they represent and that form a group of knowledge blocks that can be manipulated according to the needs of a person [3]. They represent a collection of knowledge that helps people to build their understanding of the world and to solve the problems that emerge [4].

The association between MM and learning came along the creation of David Ausubel's Meaningful Learning Theory that can be defined as learning with understanding (when people are able to use the information they have acquired to do certain jobs [5]). In order to achieve meaningful learning, the apprentice must generate a MM from a combination of previous and new knowledge that is tested through problems, and that is redefined by experience and the appropriate feedback [5].

Complex systems are hard to understand because they are organized in a multilevel way that depends on local relations that, most of the time, are not obvious. Another reason that explains the difficulty of these systems is that, due to the huge amount of events and relations that have to be processed simultaneously, a huge load of data is stored in the memory [6]. To face these troubles, MM are useful resources, because they can be seen as a small and independent piece of information, which resulted of a fragmentation process of the larger system [7].

The International Standard Organization (ISO) and the International Electronic Commission (IEC) set a definition of what being usable means [8]. The standard ISO/IEC 9241-11:1998 defines usability as *the extent to which a product can be used by specified users to achieve specified goals with effectiveness, efficiency and satisfaction in a specified context of use.* There is one problem with this definition, it gives performance criteria prominence and there are three other aspects, significant to usability that cannot be measured easily: the process of action, the consequence of the process and the emotional attachment.

MM are important to usability because they provide an insight of those aspects, letting the designers, programmers, researchers, etc. understand better what is happening inside the users' mind and, therefore, explain why users act the way they do, how they feel while using the system and how they understand it.

1.2   Online Social Networking Services (OSNS)

An OSNS can be defined as a platform that provides a private online space (user's profile) along the tools that are needed to interact with other people on the Internet allowing them to find common interests and exchange experiences and multimedia resources [9].

One of the most used OSNS is Twitter[1]. Twitter has established itself as the second largest OSNS in the world with over 500 million users[2]. Due to the dynamic nature of Twitter – the information's flow, its interactive and communication model, the new functions[3], the constant changes on the GUI and that people tend to explain it according the way they use it[4] – its complexity increases with time and optimal learning is more difficult to reach. This compels the user to update his MM frequently in order to comprehend the way it works, though some users tend to stall this effort and new users are not used to it, this causes a gap of knowledge between users with different levels of expertise.

Those factors have resulted on Twitter operating under an interactive and communication model which has not easily matched with other OSNS technologies and which demands to understand operatives which are not modeled by other forms of communication (e.g., e-mail, phone, snail mail).

The aim of this research is to identify differences among the MM that users and non-users use to understand Twitter and their impact in the participants' performance (specifically success rate and completing time) and satisfaction with the system. We aim to contribute to the HCI literature by providing more elements to understand how users learn to interact with a complex interactive system and provide valuable insights for designers of OSNS and other forms of novel communication.

2   Previous Work

One of the greatest problems of working with MM is that we only have indirect proofs of its existence, majority of the researchers infer their existence through studies involving the observation of the differences between expert and novice users or comparing the performance of users after being exposed, or not, to mental models [10].

Brandt and Uden found that people without expertise could not articulate their knowledge behind their Internet search skills; when asked why sometimes the system responded in a particular way, they tended to guess [4].

Dixon and Johnson found that one of the most important difference between novice and expert users is the capacity to comprehend the problem because novice users needed more time and effort to transfer the model made during the "understanding the problem" phase to a more familiar model used in the "solving" phase [11].

Kieras and Bovair [12] and Borgman [13] showed that people who were exposed to MM while teaching them how does a device worked and how to use it, took less time to understand the way it worked and completed the given task in a quicker and more efficient way than people that weren't exposed to the MM.

---

[1] An information network whose content is real-time updated.

[2] http://semiocast.com/publications/2012_07_30_Twitter_reaches_half_a_billion_accounts_140m_in_the_US

[3] http://techbeat.com/2013/01/is-twitter-becoming-too-complicated/

[4] http://www.rexblog.com/2010/07/19/46427

Zhang found that participants used four styles of MM to describe the Internet (Technical, Functional, Process and Connection) and that these MMs where influenced by the personality and previous experience of the participants [14].

Hmelo-Silver and Pfeffer [6] and Thatcher and Greyling [15] conducted a series of studies about the MM people defined while using complex systems (aquatic systems and the Internet). They found that participants used different styles of MM and that participants with a higher level of expertise tended to include more concepts and elaborated relations on their MM than participants with lowers levels.

Studies about how people create MM to understand OSNS or about how the sense making is done are not vast. The studies related to OSNS focus on how people use them, which new functions they can have, how they affect (psychologically) the users or how to design new interfaces to certain user groups. Some of these studies include:

Liu, Chung and Lee found that the content (information's topics) and the technological (usability and mobility) satisfaction were more important than the social (social interaction) and the process (hang out) satisfaction in order to keep people using Twitter [16].

Hargittai and Litt found that people with higher computational skills were most likely to use Twitter and, that the information gathered from a close group of people (family, friends, etc.) play a significant role in the adoption of a new technology [17].

To summarize, there have been lots of studies about MM and their relation with performance in different areas, like the Internet or biology, in order to have a better insight on the mental processes that occurred during learning, unfortunately OSNS have been omitted, since the studies about them have focused on why and how people use it.

3   Method

With an experiment conducted with 30 college students (aged from 19-25 years old), we tried to identify and compare the MM the participants defined and how they affected their performance and the perception of usability of the system.

3.1   Participants

30 participants (14 men and 16 women) were chosen randomly within our university's student community regardless their gender, semester and course. Their ages ranged between 19-25 years old ($\mu=22$, $\sigma=1.6$). They were divided into two groups (each one with 15 participants) according to their expertise with Twitter, group "Twitter" was formed by participants with a year or more of expertise and group "no Twitter" was formed by participants who were not users of Twitter.

3.2   Data Gathering and Procedure

Each participant had a 2-hour session divided in four parts: a key skills test, six Twitter tasks, a drawing activity, and a SUS test.

To assess us the knowledge participants had regarding communication technology and information management, in order to quantify possible biases, we used the Key Skills 4 U Practice Test ICT: Level 2 Test A, designed in the United Kingdom by Learning and Skills Improvement Service.

Six activities were considered as Twitter's functional core because they represent what a user must know to be considered an active user within Twitter (writing a tweet, looking for and following a user, giving retweet, using mentions, using hashtags and sending direct messages). The participants, using a Microsoft Windows (Meebox) tablet with OS Windows 8, were given a common usage context, and using Twitter's website interface, they tried to complete each task. For each one, the completing time was measured along whether the participants were successful or not.

After each activity was completed, the participants were asked to draw a MM to explain the activity they had completed and to describe it in a few sentences. After the six activities the participants were asked to draw a MM that joined all the concepts and relations regarding Twitter. For the analysis nine concepts were defined as necessary, which are: tweet, time line, follower, retweet, mentions, hashtags, trending topics, direct messages and user search

To measure users' usability perception we used one of the most used tests, the System Usability Scale (SUS). In order to be clearer while interpreting its results we used the method proposed by Bangor, Kortum and Miller [18], which associates letters to the SUS Score like schools do with grades, bearing this in mind the results were interpreted as: 90-100=A, 80-89=B, 70-79=C, 60-69=D, 0-59=F.

4   Results

4.1   Key Skills Test and SUS

A T-Test (t (22) =0.225, p<0.824; α=0.05) confirmed there was no evidence of a difference among the groups. This result is important because it assures us that the results obtained in the activities and SUS weren't biases by the lack of ICT knowledge.

The analysis showed that members from the "Twitter" group had a better perception of the system thus they graded better (grade B) the system usability than the members from the "no Twitter" group (grade C) who thought the system was complicated. Members from both groups agreed that the system was not as easy to use and to understand for the new users, therefore, they suggested some kind of tutorials to help new users understand it.

4.2   Twitter Tasks

Considering the mean of the completing times, in three activities (tweeting, using hashtags and sending direct messages; Table 1) there were no significant differences among the participants, even though participants with Twitter expertise always took less time, but in the remaining activities (searching and following, retweeting and using mentions) there were significant differences.

Table 1. Summary of the results obtained in the Twitter Tasks

Activity	T-Test (α=0.05)	Chi-Square Test
1: Tweet	t (28)= -0.87, p<0.39	1
2: Search and Follow	t (19)= -2.92, p<0.033**	0.68
3: Retweet	t (17)= -3.84, p<0.001**	1
4: Mentions	t (25)= -2.18, p<0.04**	1
5: Hashtags	t (28)= -1.18, p<0.25	0.85
6: Direct Messages	t (23)= -1.46, p<0.16	0.32

Considering the success rate, we found that none of the activities showed evidence of a significant difference among the participants. The reason that explains why there was no difference at the success rate is that participants from the "no Twitter" group after, failing to complete it at their first try, tended to use a trial-error method ·that eventually led them to the solution but it affected their completing times.

4.3   Drawing Activity

Three different MM styles were identified based on drawings and the drawing descriptions provided by the participants:

1. Analogy MM (AMM): participants relied on comparisons with other domain concepts like human activities (tweet = talk) or objects (profile = folder).
2. Technical MM (TMM): participants focused on the technological/technical elements of the system like databases, servers and computers.
3. Conceptual MM (CMM): participants only described the concepts and their relations.

The analysis showed the majority of participants used a technical view (over half of TMM's members did have a Twitter account), at the AMM there was no trend as half of its members belonged to the "Twitter" group and the other half to the "no Twitter" group, and the CMM was formed, mostly, by members from the "no Twitter group". Within the AMM and the CMM participants without Twitter expertise mentioned more of the pre-defined elements which meant they were more complete than the participants from the "Twitter" group and, TMM, was the only style in which the participants with a Twitter account used a more complete model.

5   Discussion

The main goal of this investigation was to identify the differences among the mental representations (MM) that people created while they interacted with an interactive system and their effects on the participants' performance and usability perception.

After analyzing the results we can conclude that, participants didn't have different ICT knowledge, regardless the differences in expertise with Twitter or the MM style, this was unexpected due the different characteristics the participants had like career, age and gender; this contrasted with the findings by Hargittai [17] who found that people with high computer skills were more likely to use Twitter than people with lower skills.

Regarding the Twitter activities, the analysis showed that there was a significant difference in the performance but only in 50% of the tasks, and considering the conclusions of Kieras [12], Zhang [14] and Borgman [13], this was expected because in half the tasks there was no difference no matter the level of expertise. One possible explanation for this is the tasks' complexity. The three tasks that that showed significant differences (searching and following users, retweet and mentions) can be considered as complex because they involved more steps to be done than the other tasks that had direct links or could be done in the home page. An interesting finding was that the three tasks required to search for users, so we think that the most complex task is looking for users due the GUI problems or the confusion of searching for users or words. Also, there wasn't a difference considering the success rate, although our experience was similar to Brandt's [4] due the participants without experience with Twitter tended to use more the trial-error method.

The analysis on the Drawing Activity showed that there were three styles of MM, though only one of these styles (Technical) appeared in the studies of Zhang [14] and Thatcher [15]. There was some tendencies on the MM styles, the TMM was mostly used by experienced users (eight out of thirteen) and the CMM was mostly used by non-experienced users (seven out of ten), even though we could see that there were some participants with experiences used a more complicated MM and participants with experience used a simpler MM, like Thatcher [15] we concluded that time of usage is relevant to form better MM but no necessary.

After the SUS analysis we concluded that there were significant differences on how people sees usability in Twitter. Participants with a higher level of expertise tend to see it more usable than participants with no expertise and thus they graded the system with a B. that Twitter could change in order to make it easier for new users, like better feedback, more visible buttons, etc.

Acknowledgments. We would like to thank all the participants (from the pre-tests to the final test) who gave us a couple of hours from their busy schedules. This work has been supported by Asociación Mexicana de Cultura A.C.

References

1. Moreira, M.A., Greca, I.M., Palmero, M.L.R.: Mental models and conceptual models in the teaching & learning of science. Revista Brasileira de Investigação em Educação em Ciencias 2(3), 84–96 (2002)

2. Krapas, S., Queiroz, G., Colinvaux, D., Franco, C., Alves, F.: Modelos: Uma análise de sentidos na literatura de pesquisa em ensino de ciências. Investigações em Ensino de Ciências 2(3), 185–205 (1997)
3. Johnson-Laird, P.N., Girotto, V., Legrenzi, P.: Mental models: A gentle guide for outsiders. Sistemi Intelligenti 9(68), 33 (1998)
4. Brandt, D.S., Uden, L.: Insight into mental models of novice Internet searchers. Communications of the ACM 46(7), 133–136 (2003)
5. Michael, J.A.: Mental models and meaningful learning. Journal of Veterinary Medical Education 31(1), 1–5 (2004)
6. Hmelo-Silver, C.E., Pfeffer, M.G.: Comparing expert and novice understanding of a complex system from the perspective of structures, behaviors, and func-tions. Cognitive Science 28(1), 127–138 (2004)
7. Moray, N.: Intelligent aids, mental models, and the theory of machines. International Journal of Man-Machine Studies 27(5), 619–629 (1987)
8. Bevan, N.: International standards for HCI and usability. International Journal of Human-Computer Studies 55(4), 533–552 (2001)
9. Ahn, Y.Y., Han, S., Kwak, H., Moon, S., Jeong, H.: Analysis of topological characteristics of huge online social networking services. In: Proceedings of the 16th International Conference on World Wide Web, pp. 835–844. ACM (2007)
10. Staggers, N., Norcio, A.F.: Mental models: Concepts for human-computer inter-action research. International Journal of Man-Machine Studies 38(4), 587–605 (1993)
11. Dixon, R.A., Johnson, S.D.: Experts vs. Novices: Differences in How Mental Representations are Used in Engineering Design (2011)
12. Kieras, D.E., Bovair, S.: The role of a mental model in learning to operate a device. Cognitive Science 8(3), 255–273 (1984)
13. Zhang, Y.: The influence of mental models on undergraduate students' searching behavior on the Web. Information Processing & Management 44(3), 1330–1345 (2008)
14. Borgman, C.L.: The user's mental model of an information retrieval system: An experiment on a prototype online catalog. International Journal of Man-Machine Studies 24(1), 47–64 (1986)
15. Thatcher, A., Greyling, M.: Mental models of the Internet. International Journal of Industrial Ergonomics 22(4), 299–305 (1998)
16. Liu, I.L., Cheung, C.M., Lee, M.K.: Understanding Twitter usage: What drive people continue to tweet. In: PACIS 2010 Proceedings, pp. 927–939 (2010)
17. Hargittai, E., Litt, E.: The tweet smell of celebrity success: Explaining variation in Twitter adoption among a diverse group of young adults. New Media & Society 13(5), 824–842 (2011)
18. Bangor, A., Kortum, P., Miller, J.: Determining what individual SUS scores mean: Adding an adjective rating scale. Journal of Usability Studies 4(3), 114–123 (2009)

Motivation to Self-report:
Capturing User Experiences in Field Studies

Minyou Rek, Natalia Romero, and Annemiek van Boeijen

Delft University of Technology, Faculty of Industrial Design Engineering
Landbergstraat 15, 2628 CE Delft, The Netherlands
m.b.rek@student.tudelft.nl,{n.a.romero,a.g.c.vanboeijen}@tudelft.nl
http://www.io.tudelft.nl

Abstract. User experience (UX) refers to the feelings people have when interacting with a product or service. UX design aims to enable certain experience through the development and testing of prototypes, therefore methods are needed to capture and evaluate user experience at different stages of use. Experience Sampling Method has been used to capture user experience on a moment-to-moment basis and in the context they are elicited. One mayor drawback of this method is the high load on participants, which often results in lowering participation in the study. Based on a literature review on motivational theory two design concepts are presented to illustrate how different motivators could influence different qualities of participation. Initial explorations of these concepts address opportunities and challenges of motivational mechanisms in the development of UX design and research methods.

Keywords: UX, Motivation, Self-Report, Long-term Field Studies.

1   Introduction

Momentary experiences are important to inform the design of technologies that aim to support daily life practices [2]. However capturing momentary experiences is challenged by traditional methods like questionnaires and interviews, as they rely on participants ability to recall past memories which often results in inaccurate information. Experience Sampling Method (ESM) [1] has been used in longitudinal field studies to evaluate people moment-to-moment experiences with respect to a particular situation. By prompting people several times a day to report on their experiences detailed and fine grained overview of human experiences and its variations are collected over time. Whereas recalling effects are minimized, ESM puts a high load on participants. Undesired interruptions cause annoyance, and repetitive prompts evoke feelings of burden and boredom, ultimately resulting in a negative experience for participants [3].

One way to address these issues is to minimize the aforementioned barriers. For example, minimizing interruptions by making the sampling process adaptive to participants' preferences and context of use. We argue that there are other more meaningful ways to influence participants that may result on more lasting motivation to participate. This paper reports on a User Centered Design

C. Collazos, A. Liborio, and C. Rusu (Eds.): CLIHC 2013, LNCS 8278, pp. 111–114, 2013.
© Springer International Publishing Switzerland 2013

approach that brings knowledge into the design of motivational strategies for self-report measurement tools. Based on eight motivators derived from a literature review on motivational theories two design concepts are described to explore qualities of participation that could influence the experience of self-reporting. We discuss the initial insights obtained and the implication of these interventions in the wider context of field studies are presented.

2 Design Concepts

Literature research on motivational theories resulted in a selection of biological and cognitive factors that drive people's actions. The outcomes comprises a list of 8 motivators clustered in 3 themes: *Fun* described by curiosity, surprise and joyous; *Personal Benefit* described by self-actualization/reflection, accomplishment and contribution; *Control* described by independence/autonomy and tranquility/safety.

2.1 FamilyConnector - The MailPrise

FamilyConnector is a home based awareness system that connects an older person with their adult child using touch screen displays. The aim of the system is to increase connectedness by subtle means of communication [4]. To evaluate the system, an evaluation corner based on ESM was integrated in the Family-Connector's displays to visualize the prompts which will be linked to a physical booklet to provide the answers.

MailPrise redesigned the evaluation corner and the booklet using the metaphor of a mailbox to explore qualities of control, fun and personal benefit (see Fig. 1). Digital and physical numbered envelops (orange) are used to represent time-based prompts, exploring surprise, joyous and curiosity, as prompts have to be physically search in the mailbox (fun). Digital and physical non-number envelopes (blue) are offered as free-based self-reports to bring a sense of autonomy (control). Similarly the options to accept/reject/postpone a prompt aim to provide a feeling of tranquility (control). Finally, ribbons and a mailbox filled with answered envelopes are offered as feedback to acknowledge and quantify participant's contribution (personal benefit).

2.2 ESTHER - The MoodGarden

ESTHER is a research tool designed to capture the experience of hip replacement patients when recovering at home [5]. A tablet is provided to randomly asked patients 3 times a day their mood of the moment and a short report regarding their recovery experience. At the end of the day an overall mood is asked together with a reflection of their experience of the day. The input modalities are text and voice recording using the same tablet.

The MoodGarden includes a small box with seven holes (one per day) to represent the garden and a bouquet of paper made flowers of different lengths,

Fig. 1. Left 2 pictures: MailPrise, evaluation corner and mailbox, Right 2 pictures: MoodGarden, bouquet of flowers and garden

shapes and colors, with their top wrapped and numbered to link them to different moods (see Fig. 1). After the patient reports their overall mood a message is generated with the number linked to that mood inviting the patient to pick and unwrap the flower exploring qualities of curiosity and surprise. Placing the flower in the garden provides a visualization with the purpose to evoke achievement and reflection by gradually seeing the garden getting complete, and observing daily changes of one's mood states.

3   Discussion

Both concepts are initially explored in two field study evaluations. Mailprise was deployed in a 2 weeks pilot study for the evaluation of FamilyConnector with a senior woman and her adult daughter. MoodGarden was deployed in a 1 week exploratory study using ESTHER with one THR patient (male, 70 years old) during his first week of recovery. Logging data of participants interaction with the concept and exit interviews were collected to unveil participants' experience with the concept.

Fun elements were reported by participants as strong motivators for them to continue reporting. In particular the older adult using the MailPrise reported that the envelopes on screen felt like receiving a small letter which invited her to take the moment to react to it. As reported by the participant using MoodGarden the extra effort and dedication needed to do the reporting was balanced out by the benefit of becoming a game with his grand daughter who visited him almost every evening to pick a flower and place it in the garden together. Elements related to personal benefit also seemed to influence motivation. The garden triggered the participant to reflect on his recovery progress, as it provided a memory cue about his mood states over the week. An important effect of the garden was that it drifted the attention from contributing in the researcher's goal to a personal goal (completing the garden). Control was also appreciated. The adult child reported that she accepted almost all prompts often leaving the envelope to a side to answer it later. The blue envelops allowed her to catch up with her pile of unanswered questions, at moments when she could find time to give a response, thus reducing her feeling of guilty due to missing prompts.

Rogers [6] discusses the need for new methods to support research and design practices in the complexity of real settings to bring knowledge into how people accept and adopt new technologies in their daily practices. Understanding the importance of active participation in field studies a key challenge is user motivation. Aware of the limited size and length of the interventions, these preliminary insights acknowledge the need for methods that incorporate motivational mechanism to ensure the participation on long term field studies. Several participation qualities have been explored to stimulate participation from a different perspective than traditional reward mechanisms. They serve as a starting point to look at participation like any other interaction that deserve a user centered approach to design solutions that empower participants and increase the quality of their reports.

4   Conclusion

MailPrise and MoodGarden are two concepts to explore participation qualities in ESM studies. Future work will report on the analysis of using MailPrise to evaluate FamilyConnector in a field study of four weeks with three couples. The analysis will focus on the effect of MailPrise in users' participation, as well as the quality and richness of the data obtained compare to interview techniques.

Acknowledgments. This study is funded by the research program Integral Product Creation and Realization of the Dutch Ministry of Economic Affairs.

References

1. Larson, R., Csikszentmihalyi, M.: The Experience Sampling Method. New Directions for Methodology of Social and Behavioral Science 15, 41–56 (1983)
2. Roto, V., Law, E., Vermeeren, A., Hoonhout, J.: User Experience White Paper. Dagstuhl Seminar on Demarcating User Experience, Germany (2011), http://www.allaboutux.org/files/UX-WhitePaper.pdf
3. Scollon, C., Kim Prieto, C., Diener, E.: Experience Sampling: Promises and Pitfalls, Strengths and Weaknesses. Journal of Happiness Studies 4, 5–34 (2003)
4. Vastenburg, M.H., Romero Herrera, N.A.: Experience Tags: Enriching Sensor Data in an Awareness Display for Family Caregivers. In: Keyson, D.V., et al. (eds.) AmI 2011. LNCS, vol. 7040, pp. 285–289. Springer, Heidelberg (2011)
5. Jimenez Garcia, J.C., Romero, N., Keyson, D.: Capturing patients daily life experiences after Total Hip Replacement. In: 5th International Conference on Pervasive Computing Technologies for Healthcare and Workshops, pp. 226–229. IEEE (2011)
6. Rogers, Y.: Interaction design gone wild: Striving for wild theory. Interactions 18(4), 58–62 (2011)

Crowd-Computer Interaction, A Topic in Need of a Model

Leonel Vinicio Morales Díaz[1], Laura Sanely Gaytán-Lugo[2],
Mario Alberto Moreno Rocha[3], and Adrián Catalán Santis[4]

[1] Universidad Francisco Marroquín, Guatemala
litomd@ufm.edu
[2] Facultad de Ingeniería Mecánica y Eléctrica, Universidad de Colima, Mexico
laura@ucol.mx
[3] Universidad Tecnológica de la Mixteca, Mexico
sirpeto@gmail.com
[4] Universidad Galileo, Guatemala
adriancatalan@galileo.edu

Abstract. Crowd-Computer Interaction - CCI - is a form of human-computer interaction - HCI - in which single actions from many individuals are aggregated to produce a different result that would not be achievable otherwise for one individual alone. As a research topic several questions remain open regarding CCI, for example, to what extent the principles and heuristics of interactions design under the paradigm of one-user-one-interface are applicable to crowds interacting with a network of interfaces? If a system is usable for individuals, will it be usable for crowds? Should designs be centered on the individual or on the crowd? A model of how crowds interact with computers is needed to start finding answers, that need is discussed in this paper along with some research proposals to develop that model.

Keywords: Crowd-Computer Interaction, Usability, Interaction Design, Models of Interaction.

1 Introduction

In his seminal book first published in 1895 "The Crowd: A Study of the Popular Mind" [3] the French social psychologist and sociologist Gustave Le Bon describes two concepts for crowds, the first from the ordinary sense of the word: "a gathering of individuals of whatever nationality, profession, or sex, and whatever be the chances that have brought them together." The second from a deeper analysis: "Under certain given circumstances, and only under those circumstances, an agglomeration of men presents new characteristics very different from those of the individuals composing it. The sentiments and ideas of all the persons in the gathering take one and the same direction, and their conscious personality vanishes. A collective mind is formed, doubtless transitory, but presenting very clearly defined characteristics." Le Bon recognized that what he called "psychological crowd" is different from the simple aggregation of individuals.

C. Collazos, A. Liborio, and C. Rusu (Eds.): CLIHC 2013, LNCS 8278, pp. 115–122, 2013.
© Springer International Publishing Switzerland 2013

In this paper we will argue for the need of addressing the interactions between crowds, in the sense of Le Bon's psychological crowds, with networks of computers as a different problem from that of a single user interacting with one computer, or a fairly homogeneous group of users interacting with a system, that has traditionally been the subject of HCI.

Other authors have noticed the peculiar properties of masses. James Surowiecki is the author of the book "The Wisdom of Crowds" [15] in which he argues that under conditions of diversity, independence and decentralization, collectives are smarter than individuals and even smarter than the smartest member of the group.

The emerging field of crowdsourcing [5, 12, 13] has attracted the attention of several researchers that want to find innovative ways of capitalize on the power of crowds for solving tough problems. On the behavior of individuals and crowds in a crowdsourcing setting, a technical report by a team from Microsoft Research [17] concludes that "the scale matters: individual worker behavior differs qualitatively from collective behavior."

During the CHI 2009 conference in Boston, Barry Brown, Kenton O'Hara, Tim Kindberg and Amanda Williams conducted a workshop entitled "Crowd Computer Interaction" [11] in which participants explored the possibilities of interactions between crowds and technologies designed specifically for them. Although the name of the workshop may suggest a first examination of the topic of CCI as a specific sub field of HCI we believe that it left plenty yet to be defined in order to introduce CCI as a topic of research on its own right.

What is clear is that interactions at the crowd level are different from those at the individual level. Not just because they aggregate actions but because they produce different results [3, 15]. At the individual level users work with computers to get a task done driven by particular motivations in a cyclic action process [9]. They engage differently and are rewarded differently. How well designed the interaction is becomes crucial at this level. If it is poorly designed the engagement and reward the user will get from using it will most likely be poor deriving into avoidance or reluctance to use. On the other hand, if it is well suited for the task and is easy to learn and use, enthusiasm and enjoyment would be the feelings associated with it; furthermore, happy users will share their experiences and encourage others to join the interaction [16, 18, 19].

At the crowd level the aggregated interactions can be analyzed from a different point of view. If an important number of users use a particular information artifact then several tendencies can be studied and associated with positive or negative qualities of the designed interaction. User base growth rate, average amount of time spent using, average number of outcomes by type, regularity of use, messages transmitted among users, communities being formed, and others. None of these results is expected to remain stable for long. A new cycle of aggregated interactions will change or reset the tendencies and set new ones. In that sense CCI can be said to be cyclical just as single-user HCI is.

Although intuition suggests that positive individual user experiences lead to good numbers at the crowd level, once the crowd is set and becomes the center of attention for the owners of the system, changes in the design of the interaction may start focusing on getting better tendencies at the crowd level rather than improving experiences at the individual level. Is it that under certain circumstances usability engineering switches attention from the individual to the crowd? Or should it do?

Several experiments have demonstrated that crowds are capable of performing useful tasks [1, 6, 14]. In fact there are companies whose business model consists of creating the proper environment to host a crowd and then allowing customers to hire the crowd for suitable tasks such as massively testing a particular feature in a web site, tagging pictures and discarding those that may seem offensive under a set of criteria, etc. Individuals in the crowd are rewarded with money or other incentives according to their participation in the task [2, 4, 5].

Under this business scenario the crowd is observed not only to acknowledge the trends produced in the use of the environment but also for assessing quality of results in order to make sure that customers obtain what they are paying for [2].

Several research questions remain open regarding CCI: what new aspects of human-computer interaction become apparent at the crowd level? Which ones are best studied at the crowd level than at the individual level? To what extent a successful crowd-computer interaction is related to a well designed interaction at the individual level? Is it enough to take good care of the design at the individual level to guarantee a good response at the crowd level? Are there other principles and heuristics to take into account when designing interactions for crowds? Does usability aggregates in the same form as individual interactions aggregate to produce results in CCI? What makes a user interface better for a crowd? Those questions and many others remain to be answered.

As mentioned earlier the crowdsourcing process has driven much research attention but it may be missing the perspective of seeing crowds as humans wanting to pursue their own goals when using computers and not always fulfilling them because the interaction has been poorly designed. The situation might be similar to that of early days of computers when programmers worked in a task-centered style instead of a user-centered approach.

To start finding answers to these questions a general model of crowd-computer interaction is needed. In this paper we propose such model and our plan to address the problem.

2   Characterizing Crowd-Computer Interaction

Crowd-computer interaction actually happens between a crowd and a network of computers, usually a social network. Individuals in the crowd are attracted to the network by specific stimulus they receive very much in the same way our sensory cells are stimulated by a special type of physical phenomena which they are specialized to interact with.

Individuals are like sensory cells for crowds and as with our sensory cells the result of their aggregated interactions is a completely different product. In crowds the aggregation of individual interactions produces trends and preferences some of them stable others changing at different pace in time.

The idea of individuals as cells was also proposed by Le Bon: "The psychological crowd is a provisional being formed of heterogeneous elements, which for a moment are combined, exactly as the cells which constitute a living body form by their reunion a new being which displays characteristics very different from those possessed by each of the cells singly." [3] In CCI the specification of sensory cells [20] would be better suited for the condition of cyclical information processor of each individual.

Individual interactions between a human and a computer are cyclical [9]. The individual approaches the computer with a task in mind, performs some action, obtains a result, and iterates until the result is the one desired to consider the task done. New tasks or subtasks may spark during the process triggering new rounds of interaction.

When aggregated at the crowd level, these interactions produce different types of trends over time: trends of use (intense, sparse, intermittent, etc.), trends of results, message exchanges and communications, preferences, and others. These trends could be observed from different perspectives and they can be detected at different time frames, even intertwining with each other. Each time a trend is set the crowd can be said to have completed a cycle of interaction.

Fig. 1. The cyclic nature of the interaction with computers by single users (left) is resembled by crowds using networks of devices (right)

A crucial element of CCI is the number of individuals interacting with the system [7]. If this number is low it may be regarded as not enough to produce crowd results and not to be representative of an interaction between a crowd and a network. Nevertheless defining the right number of users required to reach the threshold level and start getting crowd results remains a tricky task [12, 13].

There are several types of incentives for individuals to form an interacting crowd. Some may be explicit, known in advance to any potential user. Others are subtle, less evident and only enjoyable after some rounds of interaction [6]. In any case, once the crowd is formed there is another important incentive to join: the sense of belonging to a community [8, 21]. This one is so important that it may outweigh any other benefit.

As the interaction evolves members will gain status, credibility, notoriety, and reputation, or at least they would expect to [10]. A serious handicap in the system would be to fail to acknowledge these properties and if there is no mechanism to circumvent the failure people may feel less compelled to use it.

Finally to completely fulfill the expectations of a social network, the interaction should provide some form of triadic closure, or the ability to make friends with the friends of a friend. The presence or absence of this property can compel users to join or leave. For an interesting study on social needs and motivations in the setting of online sport communities, see [22].

3   The Elements of a Model for CCI

From all these considerations, the elements for a model of Crowd-Computer Interaction can be derived as follows:

- The number of users must exceed a certain threshold above which the crowd-computer interaction starts
- Crowds interact with networks through a multitude of platforms and devices, with varying interfaces
- Crowds interact with networks in cycles the same as individuals do with computers
- Interaction cycles for individuals produce computational results. For crowds, the results are trends they set
- The trends set by a crowd as they interact with the network are the clues to characterize the interaction
- Individuals are like sensory cells for crowds
- Individuals have to be attracted to the network by some form of incentive
- Social recognition (community belonging and gain of reputation) is a normal expectation in the members of a crowd and can be provided through mechanisms in the network
- The possibility of triadic closure (making friends with the friends of a friend) is an appealing feature to include in the interaction design.

These elements are depicted in Fig. 2. What our model proposes is that to recognize a crowd-computer interaction these elements should be assessed and that the usability of a crowd-oriented application relies on their appropriate adjustment after observing the trends outputted in an iterative process much in the same way the usability of user interfaces designed for single users is tweaked observing reactions of users to prototype changes.

4   Validating and Using the Model

To validate the model it has to be tested, and possibly adjusted, against as many crowd-computer interactions as possible.

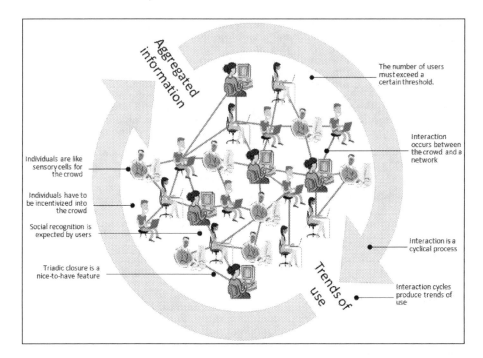

Fig. 2. The cycle of interaction between crowds and computers and the elements of the model

Because there is no ready-made list of examples of crowd-computer interactions a set will be compiled based on the criteria we had at the beginning of the project: the subject of study is interactive systems where crowds of users produce results different from and beyond those of individual interactions.

Checking the model against examples is the first stage of validation.

For a second phase we plan a crowd-computer interaction experiment. An online form (a single user interface deployable to many) is to be shared with as many knowledgeable people in the field of HCI (the crowd) as possible to request examples of CCI congruent or not with the proposed model. The form can be filled as many times as needed (allowing iteration) and the results updated and shared back as often as possible with the names of contributors visible (to provide social recognition). Our team has already started this process, in a limited version, with good results and interest from the community.

Finally, after some iteration, the form will be closed and a list of contributors will be published along with the results of this project. This incentive will be announced in the invitation to participate so it serves as a perceived benefit (the stimulus to join).

The experiment is expected to yield examples that match the model as well as others that need explanation or point to adjustments in the model. Because this is only a work in progress we are not completely sure of how the proposed model could be used or with what aim. One of the expectations of the team is to provide the basis for usability engineering at the crowd-computer interaction level, including equivalents for prototype testing, heuristic evaluation, user testing, and other techniques.

Our first intuition is that the equivalent of user reaction at the crowd level can only be trends set in use. A usability evaluator at the single user level pays attention to facial, body and verbal expressions, struggles, indications, thinking aloud verbalizations, and other clues for determining adequacy between design and intended users for the tasks to be performed with the interface. At the crowd level the trends that emerge when considered from different points of view perform that function.

5   Conclusion

In this paper we have explained why the phenomena related to Crowd-Computer Interaction – CCI – deserves especial attention from the HCI research community and how this attention could be delivered.

In is important to note that although the topic of crowdsourcing is being studied abundantly as shown in the several references included here, there are important characteristics of crowds that need to be addressed according to their peculiar nature especially when crowds interact with computers through networks. Crowdsourcing should not be considered completely equivalent to CCI. When considering crowdsourcing and CCI a parallelism could be made with the duet of computers as productivity tools and HCI as the study of humans using computers. Computers and software are powerful tools to solve problems, and so is crowdsourcing, but if their use is designed without considering the insights provided by HCI and CCI the results can be much less than optimal.

Several research questions where proposed in the paper. To start searching for answers an incipient model of crowd-computer interaction was presented. It is the model of a cyclic process with properties that are considered desirable. Our plans to test and validate the model where also shared. The future work is implicit.

References

1. Bozzon, A., Brambilla, M., Ceri, S.: Answering search queries with CrowdSearcher. In: Proceedings of the 21st International Conference on World Wide Web (WWW 2012), pp. 1009–1018. ACM, New York (2012)
2. Hoffmann, L.: Crowd control. Commun. ACM 52(3), 16–17 (2009)
3. Le Bon, G.: The crowd: A study of the popular mind. Macmillan (1897)
4. Doan, A., Ramakrishnan, R., Halevy, A.Y.: Crowdsourcing systems on the World-Wide Web. Commun. ACM 54(4), 86–96 (2011)
5. Greengard, S.: Following the crowd. Commun. ACM 54(2), 20–22 (2011)
6. Kazemi, L., Shahabi, C.: GeoCrowd: Enabling query answering with spatial crowdsourcing. In: 20th International Conference on Advances in Geographic Information Systems (SIGSPATIAL 2012), pp. 189–198. ACM, New York (2012)
7. Liang, Y., Caverlee, J., Cheng, Z., Kamath, K.Y.: How big is the crowd?: Event and location based population modeling in social media. In: Proceedings of the 24th ACM Conference on Hypertext and Social Media (HT 2013), pp. 99–108. ACM, New York (2013)

8. Haythornthwaite, C.: Learning networks, crowds and communities. In: Proceedings of the 1st International Conference on Learning Analytics and Knowledge (LAK 2011), pp. 18–22. ACM, New York (2011)
9. Norman, D.A.: Stages and levels in human-machine interaction. Int. J. Man-Mach. Stud. 21(4), 365–375 (1984)
10. Bozzon, A., Brambilla, M., Ceri, S., Mauri, A.: Reactive crowdsourcing. In: Proceedings of the 22nd International Conference on World Wide Web (WWW 2013). International World Wide Web Conferences Steering Committee, pp. 153–164. Republic and Canton of Geneva, Switzerland (2013)
11. Brown, B., O'Hara, K., Kindberg, T., Williams, A.: Crowd computer interaction. In: CHI 009 Extended Abstracts on Human Factors in Computing Systems (CHI EA 2009), pp. 4755–4758. ACM, New York (2009)
12. Roughton, A., Downs, J., Plimmer, B., Warren, I.: The crowd in the cloud: Moving beyond traditional boundaries for large scale experiences in the cloud. In: Lutteroth, C., Shen, H. (eds.) Proceedings of the Twelfth Australasian User Interface Conference (AUIC 2011), vol. 117, pp. 29–38. Australian Computer Society, Inc., Darlinghurst (2011)
13. Estellés-Arolas, E., González-Ladrón-De-Guevara, F.: Towards an integrated crowdsourcing definition. J. Inf. Sci. 38(2), 189–200 (2012)
14. Bernstein, M.S., Little, G., Miller, R.C., Hartmann, B., Ackerman, M.S., Karger, D.R., Crowell, D., Panovich, K.: Soylent: A word processor with a crowd inside. In: Proceedings of the 23rd Annual ACM Symposium on User Interface Software and Technology (UIST 2010), pp. 313–322. ACM, New York (2010)
15. Surowiecki, J.: The wisdom of crowds. Anchor (2005)
16. Nielsen, J.: Usability Engineering. Morgan Kaufmann Publishers Inc., San Francisco (1995)
17. DiPalantino, D., Karagiannis, T., Vojnovic, M.: Individual and collective user behavior in crowdsourcing services. Technical report, Microsoft Research (2011)
18. Rogers, Y., Sharp, H., Preece, J.: Interaction design: Beyond human-computer interaction. Wiley (2011)
19. Rosenberg, D.: The myths of usability ROI. Interactions 11(5), 22–29 (2004)
20. Sensory Receptor Cells – MeSH – NCBI, http://www.ncbi.nlm.nih.gov/mesh?Db=mesh&term=Sensory+Receptor+Cells
21. Blanchard, A.L., Markus, M.L.: Sense of virtual community - maintaining the experience of belonging. In: Proceedings of the 35th Annual Hawaii International Conference on System Sciences, HICSS, January 7-10, pp. 3566–3575 (2002)
22. Ojala, J., Saarela, J.: Understanding social needs and motivations to share data in online sports communities. In: Proceedings of the 14th International Academic MindTrek Conference: Envisioning Future Media Environments (MindTrek 2010), pp. 95–102. ACM, New York (2010)

Effectiveness Measurement Framework for Field-Based Experiments Focused on Android Devices

Ivan Pretel and Ana B. Lago

Deusto Institute of Technology - DeustoTech
MORElab – Envisioning Future Internet
University of Deusto, Avda. Universidades 24, 48007 - Bilbao, Spain
{ivan.pretel,anabelen.lago}@deusto.es

Abstract. Most of the mobile phones have turned into full-connected devices. This provides companies with a perfect channel to interact with their potential clients and employees. The quality of the experience with these applications can directly affect the profits of the company it represents. Focusing on the mobile field and its extremely dynamic context, the quality of the experience can highly fluctuate. Inside this field, several methods and tools have been developed by defining a context of use. However, current methods can only capture it through adding external capture tools (added cameras, human observers...) that can change the experience. The main contribution in this article is a new approach to automatically measure effectiveness through a tiny but powerful mobile tool that can capture interaction metrics and the surrounding context without biasing the measured experience.

Keywords: Mobile HCI, evaluation, effectiveness, context, quality, usability, mobile services, framework.

1 Introduction

In the last few years mobile devices are gaining more and more importance to perform tasks not only in our leisure time but also at work. Companies are progressively increasing the number of services connected to the virtual world through the mobile devices. Thanks to these devices they can expose their business models to everywhere by developing and deploying a tiny mobile application. The main aim of this kind of applications is to enable potential users to interact with the business model of the company from everywhere.

Testing tools have drastically changed and have been focusing on the web domain. However, software applications must be focused not only on the web domain but also on mobile devices. According to the last Cisco Visual Networking Index[1], the average smartphone usage has nearly tripled in 2011 and the

---

[1] Cisco Visual Networking Index: Forecast and Methodology, 2011-2016.
http://www.cisco.com/en/US/solutions/collateral/ns341/ns525/ns537/ns705/ns827/white_paper_c11-481360.pdf

C. Collazos, A. Liborio, and C. Rusu (Eds.): CLIHC 2013, LNCS 8278, pp. 123–130, 2013.

average amount of traffic per smartphone in 2011 was 150 MB per month, up from 55 MB per month in 2010. Paying attention to this report, software testing tools should evolve into web and mobile at the same time.

The quality of mobile applications can extremely fluctuate depending on the context. Therefore, capture information without biasing the context is very important.

In this article we expose a mobile-based tool to automatically evaluate the effectiveness of interactions and capture metrics of the surrounding context. It is formed by a tiny Android library developed for mobile applications used to log interactions and a context model. It is also formed by a web server to remotely store all information. A preliminary version of the system which can capture the majority of the mobile context model was developed and validated through a tiny mobile game.

Firstly, the main definitions of effectiveness, usability and quality based on standards are studied in Section 2. Secondly, a context model focused on mobile devices is explained and a new approach to capture it is defined through the Section 3. In Section 4, the existing systems to capture the context model are studied and the capture tool is presented. After describing it, a brief experiment and its results are shown in Section 5. Finally, the research is concluded and further work is discussed in Section 6.

2   Usability, Quality and Effectiveness

According to the ISO 9241-11 standard [3] usability is the extent to which a product can be used by specified users to achieve specified goals with effectiveness, efficiency and satisfaction in a specified context of use. This standard defines effectiveness as the level of accuracy and completeness with which users achieve specified goals. As we will see later, it uses the same definition provided in ISO 9126-4 [4] to define the effectiveness but does not provide a general rule for how measures should be chosen or combined. In fact, it delegates the responsibility for developing the proper metrics to the product developers. This is because the importance of components of usability depends on the context of use and the software which is going to be tested. According to the ISO/IEC 9126 standard, quality represents a property of the software product defined in terms of a set of interdependent attributes (such as usability, security, reliability, performance, complexity, readability, reusability) expressed at different levels of detail and also taken into account the particular context of software use.

The different attributes can measure three different quality aspects: Internal Quality, External Quality and Quality in Use. Internal Quality is the totality of attributes of the software product from an internal view (e.g. spent resources, analysability). It is measured and improved during the code implementation, reviewing and testing. External Quality is the quality when software is running in terms of its behaviour (e.g. number of wrong expected reactions).

Quality in use is the quality of software that user can perceive when the software is used in an explicit context of use. It measures the extent to which users can complete their tasks in a particular environment. It is measured by four main capabilities of the software product in a specified context of use: effectiveness, satisfaction, productivity and safety.

Each capability is made up by several metrics which can be measured through designing and performing experiments. This work is centred on measuring the effectiveness inside the quality in use aspect, where the effectiveness is formed by three main metrics (see Table 1): Task Effectiveness (TE), Task Completion (TC) and Error Frequency (EF). These metrics measure the accuracy and completeness with which goals can be achieved.

Table 1. Effectiveness metrics defined by ISO/IEC 9126

Metric	Formula	Definition
Task Effectiveness (TE)	$\|1 - \sum Ai\|$	What proportion of the goals is achieved?
Task Completion (TC)	TCM/TA	What proportion of the tasks is completed?
Error Frequency (EF)	E/T	What is the frequency of errors?

Task Effectiveness measures the quantity of the goals achieved by a user. It is measured summing the number of errors (Ai) appeared during the task. Many errors could be more important than others. In order to solve so, each kind of error has its associate weight. TEs value will be between 0 and 1, the closer to 1 the better.

Task Completion measures the level of success the user achieves performing tasks. In contrast to the Task Effectiveness, this metric assumes that the tasks can be performed without the chance of being partially completed. In this case, it is calculated by the number of tasks completed (TCM) divided by the number of tasks attempted (TA). TCs value will be between 0 and 1, the closer to 1 the better.

Finally, Error Frequency measures the number of times that an error is made within a given period. It is calculated dividing the number of errors (E) by the task time or the number of total steps (T). This metric is very useful for making comparisons if errors have equal importance, or are weighted. EFs value will be between 0 and 1, the closer to 0 the better.

Centring on the usability and quality in use fields, effectiveness does not take account how the goals were achieved, only the extent to which they were achieved. Effectiveness, which defines and forms usability and quality in use, can drastically affect to the quality and usability. Furthermore, the defined effectiveness may be biased by the surrounding context of use.

3   Context Model Focused on Mobile Devices

Throughout the different definitions of effectiveness the context of use has been appearing as an element which can bias effectiveness, usability and quality. Many decades ago the context of use has been taken into account and has been defined several times by a lot of researchers, experts and communities.

Several studies maintain that a context is just the physical location [10]. Others add to the location more attributes such as weather [2] to achieve a more accurate definition of the environment and physical context. Other studies expand the limit of the context of use adding the community [6] and the stakeholders [7]. They also maintain that context should be defined by answering where the user is, who the user is with, and what resources are nearby. Others [9] define the context by enumerating more parameters such as goals, attention, connection

After seeing the different definitions of contexts and paying special attention to the work by Abowd et al. [1] we have defined a preliminary context of use model focused on the mobile field. We conclude that components which define the context (see Fig. 1) are: the user, the mobile device and the environment (physical, ambient, technical and sociocultural). User is defined by four main groups of attributes: personal information, knowledge, skills and attitudes. The mobile device is formed by six groups: connections, body, inputs, outputs, battery and software features. Finally, the environment is made by four main groups: physical, ambient, technical and sociocultural groups.

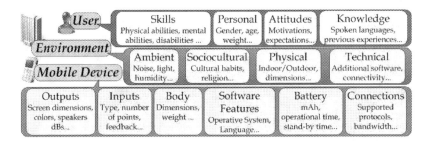

Fig. 1. Components which define the context

4   Context Model Capturer

Once the context of use model is defined the next step is to design the capturing method. To capture the context model and the interactions, it is essential to capture objective information using the mobile device in real environments. This can hardly be captured with a lab-based framework (such as Morae Observer[2]) which logs information in a highly controlled environment using specific devices and users. The field-based evaluation frameworks [8] [5] can provide more objective information because they are performed in real environments, but the added agents such as cameras and invasive evaluation methods (e.g. surveys during tasks) have to be removed. Therefore, the best way to capture interaction data and the context model is by registering information through a mobile device using a tiny capture tool. So as to implement a capturer without biasing

---

[2] Morae Observer - http://www.techsmith.com/morae.html

the context we have to use only elements that make up the context (in this case, the mobile device). This tool should capture the context model via the built-in mobile sensors and logging interaction events. Although the context model has been defined, the preliminary version only can capture a small subset of data and assumes all the errors have the same relevance. The purposed system (see Fig. 2) is formed by a tiny Android library developed for mobile applications and a server to store and log the performed interactions. First, users should download an application-to-test (ATT) from the server. They sign up for the platform, use the application and upload their interactions to the platform. The ATT should be integrated with the developed library. This library automatically captures context and interaction information through a context model module and stores it in a local database using an interaction store module. When the device has internet connection and its owner wants it, all the information is uploaded.

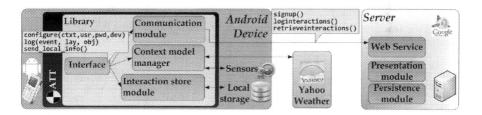

Fig. 2. Context Model Capturer Architecture

4.1 Server

The server is developed by the Google App Engine technology[3]. Its main aim is to be the main gate between users and experimenters. It remotely stores all the interaction data in the cloud. It offers several services through a web interface. Through these services users can sign up, log interactions and see their logged interactions. All the information is stored in the cloud by Google Cloud SQL[4].

4.2 Android Library

In order to capture information from the context model this library allows programmers to abstract the capture. Developers only need to implement their own application and insert small log lines inside their ATT logic to log interactions.

This library only exposes three instructions. The CONFIGURE instruction sets up the capturer with the needed parameters to work. It does not start the capture, only sets the username, pass, registered id of the device and the context (interface provided by Android to see the global information about the application environment). SEND_LOCAL_INFO sends the captured and locally stored

---

[3] Google App Engine: build and host web application -
http://developers.google.com/appengine

[4] Google Cloud SQL - http://cloud.google.com

information to the server. LOG captures the interaction timestamp, context and the object with witch user is interacting.

A task (see Fig. 3) can pass through four main states: When a task is not started yet (NOT STARTED), when a task is started and its user is interacting to achieve the goal of the task (STARTED), when a task is started but its user is not interacting to achieve it (PAUSED) and when the task is finally ended (ENDED). During a task performance a user can trigger two main events: START_TASK and END_TASK. Additionally there exist two others: if user leaves the task (e.g. phone call) the PAUSE_TASK event is triggered. Where user decides to continue the task RESUME_TASK is produced. When a task is started two events related to the interaction of the user can be triggered. The INTERACTION event means that a user is interacting in the right way. This event should be triggered when a user is achieving little microchallenges inside the goal of the task. The ERROR event means that a user has made a mistake.

1.EVENT_TYPE_START_TASK
2.EVENT_TYPE_PAUSE_TASK
3.EVENT_TYPE_RESUME_TASK

4.EVENT_TYPE_INTERACTION
5.EVENT_TYPE_ERROR
6.EVENT_TYPE_END_TASK

Fig. 3. States of a task and its events

The context model is captured by using all the built-in sensors and the application programming interfaces (APIs) provided by Android. The information related to the user is retrieved during the signup because the changing frequency of this information is very low. The nickname, gender, birthday, height, weight, if the user is left-handed, right-handed or ambidextrous and the European level of several languages are stored. Mobile device information is captured in two different phases: during the sign up (DeviceID, OS version, productID, display model, country and its configured language) and during the task performance. When the user is performing tasks information related to the sound level of outputs is captured (i.e. alarm volume, ringtone volume, if headphones are used). Information related to the battery (battery level, charging/unplugged) as well as the application display properties (density, height, and width) are also captured. Information related to the environment is captured during the task performance because of the high frequency of changes. Noise and light levels are captured by sensors. Location and connection information (coordinates, the connection type...) is captured by internal services provided by Android. If the device is connected the current weather conditions are requested to Yahoo Weather[5]).

5   System Validation

The preliminary version of the system was validated performing a brief experiment. 4 subjects have been signed up from the platform and have played a

---

[5] Yahoo! Weather Developer Network - `http://developer.yahoo.com/weather/`

memory game application which contains the library. They have been playing during one day and they have generated more than 1400 logged lines. They played in four contexts: at home (H), walking down the street (S), travelling (PT) by a public transport and at work (W), more concretely, in an office.

First, users have to sign up on the testing platform and complete information related to them and their device. Finally, user presses play button to start the game, selects a context name(H, S, PT or W) and the game starts. It is worth taking into consideration that users must manually select the context name. This is because the framework captures the conforming context variables so as to analyse how they affect to the experience, and thus it is not its aim to address the context inference issue. The memory game application is a card game in which a player deals out a set of cards face down. In one turn, the player flips over two cards (one interaction). If they match, the player leaves them face up and solves a simple add to continue playing (it is one interaction). If they do not match (error), the player flips the cards back face down.

Focusing on the effectiveness measurement, the task to perform is clear: to end up with all of the cards flipped face up in less than 15 turns. There are 8 pairs of cards; it means the best round is made by only 8 interactions and 0 errors. If a user spends more than 15 turns, it loses and this task is not completed. Task Completion (TC) is calculated counting all the rounds won by a subject divided by all rounds this subject has played. The Task Effectiveness (TE) metric is calculated subtracting 1 to the number of made errors during the task multiplied by the weight of the error. The weight of an error is $1/7$ because the maximum number of errors you can make is 6 with 8 correct interactions. If you make 7 errors the TC should be 0 $(1 \ (7*1/7) = 0)$. Error Frequency (EF) is calculated dividing the error number by the number of total turns.

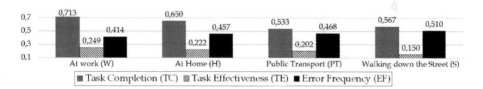

Fig. 4. Effectiveness results grouped by contexts

The results grouped by contexts (see Fig. 4) shows the outdoor contexts (PT and S) affect subjects and lead them to be less effective than in the indoor contexts (H and W). TE indicates that the S context is the context where subjects have the worst results. Although EF exposes that is in W where subjects make more mistakes, it is also the most efficient context.

6 Conclusion

Through this work we have defined effectiveness metrics, which can bias the usability and the quality of mobile applications. We have also studied context models and we have defined one based on mobile devices. This context model is formed by several groups of attributes which have to be captured and logged. Based on the study of the lab-based and field-based capturing methods and focusing the design on mobile environments we conclude the best way to capture interaction data and the defined context model is through the used mobile device.

The purposed tool is formed by a tiny Android library used to log interactions and the context model and a server to store all the captured data. Finally, a preliminary version of the system which can capture the majority of the context model and a tiny application to test it were developed. The results of the exposed experiment demonstrate the effectiveness and the context can be automatically measured by an automatic tool without biasing the interaction with external agents. The next step is to study all the captured attributes of the context model to calculate correlations with the effectiveness as well as keep enhancing the captured attributes and the system performance.

References

1. Abowd, G.D., Dey, A.K., Brown, P.J., Davies, N., Smith, M., Steggles, P.: Towards a better understanding of context and context-awareness. In: Gellersen, H.-W. (ed.) HUC 1999. LNCS, vol. 1707, pp. 304–307. Springer, Heidelberg (1999)
2. Brown, P.J., Bovey, J.D., Chen, X.: Context-aware applications: From the laboratory to the marketplace. IEEE Personal Communications 4(5), 58–64 (1997)
3. ISO (ed.): ISO 9241-11:1998(E): Ergonomic requirements for office work with visual display terminals (VDTs) Part 11: Guidance on usability (1998)
4. ISO (ed.): ISO/IEC TR 9126-4:2004(E): Software engineering Product quality Part 4: Quality in use metrics (2004)
5. Jensen, K.L.: Recon: Capturing mobile and ubiquitous interaction in real contexts. In: Proceedings of the 11th International Conference on Human-Computer Interaction with Mobile Devices and Services, p. 76. ACM (2009)
6. Kankainen, A., et al.: Thinking model and tools for understanding user experience related to information appliance product concepts. Helsinki University of Technology (2002)
7. National Institute of Standards and Technology: Common industry specification for usability requirements (NISTIR 7432) (June 2007)
8. Raento, M., Oulasvirta, A., Petit, R., Toivonen, H.: Contextphone: A prototyping platform for context-aware mobile applications. IEEE Pervasive Computing 4(2), 51–59 (2005)
9. Ryan, N.S., Pascoe, J., Morse, D.R.: Enhanced reality fieldwork: The context-aware archaeological assistant. In: Computer Applications in Archaeology (1998)
10. Schilit, B., Theimer, M.: Disseminating active map information to mobile hosts. IEEE Network 8(5), 22–32 (1994)

Evaluation of a Driving Simulator with a Visual and Auditory Interface

Juan Michel García-Díaz[1,*], Miguel A. García-Ruiz[2],
Raúl Aquino-Santos[1], and Arthur Edwards-Block[1]

[1] College of Telematics, University of Colima, Avenida Universidad 333,
C.P. 28017, Colima, Mexico
{aquinor,arted}@ucol.mx
[2] Department of Computer Science and Mathematics,
Algoma University / Sault Ste. Marie, Ontario, Canadá
miguel.garcia@algomau.ca
jgdiaz@ucol.mx

Abstract. Millions of driving accidents occur worldwide each year causing more than a million fatalities. Although traditional safety measures are largely reactive in nature, the application of wireless technologies has become much more common, thus promoting proactive strategies to save lives. This article presents the development and evaluation of usability of a driving simulator with a visual and auditory interface to assist drivers more quickly identify emergencies on the road, which, when used with the support of wireless ad hoc networking, can contribute to reducing vehicular accidents. The usability results obtained in this study were favorable according to the System Usability Scale (SUS) usability questionnaire, which was applied as a post-test. Employing the SUS, respondents reported the interface to be acceptable or good. Results show that utility of the visual interface was 69% and the score for the auditory interface was 100%. In sum, respondents felt the interfaces were useful in reported upcoming emergency or accident situations.

Keywords: driving simulator; auditory interface, visual interface, driver distraction.

1   Introduction

Vehicle transport is part of people's daily lives as it is the primary mode of transportation used by people as they carry out their daily activities. Despite advances in the area of vehicular safety, there are still many areas of opportunity as the loss of life and property is still staggering.

According to data presented by the World Health Organization (WHO), each year, worldwide, countries lose 1-3% of their GDP in traffic-related incidents. More importantly, however, between 20 and 50 million people are injured and approximately 1.3

---

* Corresponding author.

C. Collazos, A. Liborio, and C. Rusu (Eds.): CLIHC 2013, LNCS 8278, pp. 131–139, 2013.

million people die, which can be translated into one person dying every 24 seconds due to a traffic accident, making traffic accidents one of the 10 leading causes of death. If this trend continues unabated, by 2030, traffic accidents will become the fifth leading cause of death worldwide [1].

Driver assistance technologies have been proposed as a useful alternative to reduce traffic accidents and increase safety (excellent reason to implement). Their main idea is to provide the necessary information to drivers to help them make timely decisions when facing emergency situations and avoid driver distraction [2]. Automobile companies are currently integrating much of this technology into their vehicles and are conducting further research into expanding its use.

This paper presents a driving simulator integrated with a visual and auditory interface. The auditory interface produces two alarms with audio in the AT & T Natural Voices ® Text-to-Speech Demo [3]. The voice produces a caution and danger message by means of two computer speakers using both male and female voices. Past research has provided many compelling examples of TTS interfaces working as auditory warning systems (e.g. [4]). There are various examples concerning their application in in-vehicular human-computer interfaces, including work done by [5], which analyzed the use of synthesized male and female voices for auditory warnings. However, research has focused on the design and usability of in-vehicular TTS interfaces.

The visual interface consists of an electronic circuit that generates an alarm via LED (stands for Light-Emitting Diode) flashes, using an LED ultra-bright yellow that represents a caution and an LED ultra-bright red that represents a danger.

The following section of this paper explores related work. Section 3 explains how system is evaluated, describing the driving simulator, the audio and the visual interfaces, the participants and the experimental testing procedure. Section 4 discusses the results and their interpretation. The final section of this paper then provides conclusions and offers suggestions for future work.

2   Related Projects

The core of our research corresponds to the development of an auditory and visual interface and its usability testing, which was used by participants who used a driving simulator. The simulator presented in this work was similar to the one developed by Sodnik et. al. (2007, 2008), consisting of a Logitech MOMO Racing module with gas, clutch and brake pedals, along with a stick shift and a steering wheel. For the purpose of this study, we also used a, a 2.4m x 1.8m projection screen and a 7.1-channel sound and RACER software version 2.1. Finally, the interfaces used by Sodnik et. al. consist of a small screen, a Nokia Series 60, and a speaker [6, 7].

Garzon (2012) also equipped a driving simulator using a game kit that includes a steering wheel, a stick shift, gas, clutch and brake pedals, a central control unit, a racing game in 3D, and a computer with two screens. One screen is used as an interface showing a website that measures capacity features, such as time and fuel level [8].

On the other hand, Man Ho et. al. (2010) presents a driving simulator consisting of a digital projector, a controller screen with a 2400 x 1800 mm resolution, STISM Drive software (Technology Systems, Inc.) to provide images of the road and the car, a computer, and a real car (Smart, Mercedes-Benz). They used expert sound to create an interface with 70 different types of warnings generated by varying the frequency, duration and intensity [9] of the sound output.

To capture the attention of the driver, Cuong et. al. (2012) used both visual and auditory interfaces, as well as the combination of both. His driving tests were conducted in two different scenarios; the first uses a real-world car and the second one employs a driving simulator [10]. In addition, attention on in-vehicular technologies has been extensively researched over the past two decades (e.g. [11, 12]).

3   Evaluation

This work employed a usability evaluation to measure the ease of use and acceptability of the driving simulator. Usability is a set of qualitative and quantitative metrics that measure how effective, efficient and satisfactory the user experience is for persons employing a human-computer or human-digital product. One important aspect considered by usability evaluations is the interface's ease of use [13]. The instrument used to measure usability in this work is the System Usability Scale questionnaire. This questionnaire has been used with great success for many years to measure the usability of digital products and software systems worldwide [14]. The SUS consists of 10 questions that employ a Likert scale (1. "Strongly Disagree", 2. "Disagree", 3. "Neutral", 4. "Agree" and 5. "Strongly Agree") whose odd questions are developed positively, while even questions are written in the negative. Importantly, the SUS questionnaire provides a usability score from 0 (null usability) to 100 (very high usability). In this study, each participant performed the driving experience, completed a demographic background questionnaire, and completed SUS and other questionnaires that were developed expressly to evaluate the usability of the visual and auditory interface.

3.1   Materials

Vehicle control in the simulation are performed with a stick shift, gas, clutch and brake pedals, and the steering wheel that are included in the Logitech G27 Racing Wheel [15] simulator. Additionally, other components employed in the simulation included: a computer, projector and a 2.4 mx 1.8 m projection screen. The RACER 3D software version 0.8.35 [16] is used for the actual driving simulation. The Lower Class 1 level was chosen to provide easier handling and the track selected was the A-1 Ring Austria 2001 (Figure 1). These options were chosen because they are easier to work with while providing both straightaway and curve conditions.

Fig. 1. Driving Simulator with a visual and auditory interface.

The auditory interface used in this study employs two auditory alarms using verbal messages with a male and female voice to relay the following messages: "Danger, accident at 200 meters!" and "Caution, highway construction at 200 meters!". The audio was generated by the AT&T Natural Voices ® Text-to-Speech Demo software and reproduced by Dell AX210 USB Stereo Speaker System.

The visual interface representing the message "Danger, accident at 200 meters!" is provided by an ultra-bright red LED light while an ultra-bright yellow LED light represents the message "Caution, highway construction at 200 meters!". Both LEDs blink intermittently when the alarm is activated. The schematic diagram shown in Figure 2 shows an arrangement of components that comprise the electronic circuit allowing the LED lights to flash.

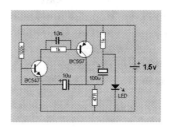

Fig. 2. Schematic diagram of the visual interface[17].

3.2   Participants

Usability testing was conducted with a group of 12 students of the Masters in Computer Science at the University of Colima and computer technicians of Siteldi Solutions, a small business dedicated to innovation and technological development, located in the City of Colima. All of the participants reported having knowledge of how to drive a car and had an average experience of 7.5 years. The average age of the participants was 26 years. As far as gender is concerned, 17% were female and 83% were male. This significant gender difference is due to the very biased male-female ratio of students choosing to study engineering degrees in Mexico, especially in graduate degree programs. Of the total universe of participants, 25% reported having previous-ly used a driving simulator and 83% reported having previously played at least one 3D video game before participating in the study. Of the participants previously having played a 3D game, 8% reported playing 3D games frequently, 57% reported playing 3D games occasionally and 17% reported playing them rarely.

3.3   Procedure of the Experimental Tests

The tests were carried out in four stages (Figure 3), in which all participants collabo-rated in a voluntary and individual manner. The first stage consisted of obtaining consent from the participants. Information was provided concerning vehicle safety and regarding the use of the driving simulator. They were then introduced to the two types of messages to be used as part of the interface to be tested in this study. In one of the messages, drivers were expected to be cautious and reduce their speed; the second message indicated a dangerous situation, which might force them to pull off the road or momentarily stop. Additionally, participants were provided five minutes to familiarize themselves with the use of the driving simulator. In the second stage, par-ticipants were asked to drive a car within the simulated environment uninterrupted in order for them to gain experience in handling the vehicle in a natural setting without hazards. After this, the participants drove over the same simulated course and each of the two auditory alarms was randomly repeated 3 times. At the end of their simulated driving experience, a questionnaire to assess the auditory interface was given partici-pants to complete. The participants were then given a short break before proceeding to the third stage, which was identical to the second stage, but focused on testing the visual alarms. Finally, the fourth part of the study consisted of applying the SUS Usa-bility Questionnaire to the participants to evaluate their experience on the driving simulator.

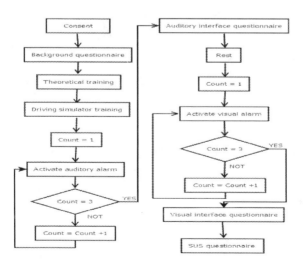

Fig. 3. Experimental testing procedure

4   Results and Taking on a Relative Interpretation

4.1   SUS Questionnaire

The SUS has proven to be a simple, effective and accurate questionnaire to assess usability [14], making it a widely used instrument to evaluate hardware, websites, mobile phones, and interactive systems voice response, among others [18].

Once results are obtained, the SUS questionnaire provides a usability value for the digital product, which can range from 0 (null usability) to 100 (very high usability). To evaluate the value obtained one must interpret the results, which implies taking a numerical value and converting into an adjective that provides a relative value. Bangor et. al. (2008.2009) has added seven adjectives associated with the Likert scale and three levels of acceptability to help improve the interpretation of scores obtained from SUS questionnaire. Table 1 shows how to suggest interpret the results.

Table 1. SUS scores with their corresponding adjective [19] and acceptability [20] ratings

SUS Scores	Adjective Ratings	Acceptability
89~100	Best imaginable	
84~88	Excellent	Acceptable
71~83	Good	
50~70	OK	Marginal
32~49	Poor	
20~31	Awful	Unacceptable
0~19	Worst imaginable	

Each SUS questionnaire scored and calculated the average assessment of the participants; the final average SUS score was 76. Table 1 provides the SUS scores with their adjectives to more adequately provide results. The results show that the driving simulator has a "good" and "acceptable" level of usability. Likewise, results show similar results for individuals as all participants reported usability above 60 points.

4.2   Assessment Questionnaire of the Visual and Auditory Interfaces

Participants also answered a 10-item questionnaire to assess the auditory interface (auditory alarms) and a 7-item questionnaire to assess the visual interface (visual alarms). A 5-point Likert scale was used with the number 1 representing "Strongly Disagree" and 5 representing "Strongly Agree". The sum of the "Agree" and "Strongly Agree" responses were then calculated. The percentage results are shown in Figures 4 y 5.

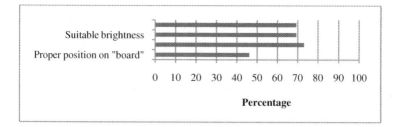

Fig. 4. Aspects evaluated in the visual interface

In relation to the results of the visual interface (Figure 4), 69% of the respondents believe the visual alarm was useful in representing the warning and 46% opined that the position of visual alarms on the "board" of the simulator was adequate. These combined results present an interesting opportunity area for future work. Two additional aspects were then evaluated: visual attractiveness and suitable brightness, where 73% responded that the visual alarm was pleasing to the eye and 69% said the brightness of the visual alarm was adequate.

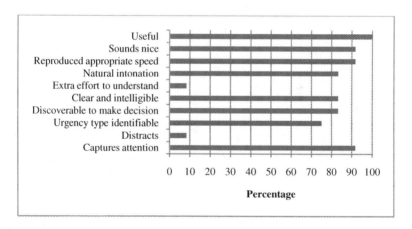

Fig. 5. Aspects evaluated in the auditory interface

Regarding how easy it was to use the auditory interface (Figure 5), 8% opined that it was necessary to make an extra effort to understand, 75% reported that they easily identified the type of emergency the auditory alarms represented, 83% easily recognized the meaning and action required for each of the auditory alarms. Insofar as the perception of the usefulness is concerned, 100% of the participants felt that the auditory alarms helped forewarn them of an emergency situation in the simulation. Furthermore, 92% reported that the auditory alarms captured their attention and only 8% indicated that the auditory alarms caused distraction. With respect to the four remaining aspects that were evaluated, 83% believe that auditory alarms were heard clearly and intelligibly, 83% stated auditory alarms possessed natural voice inflections,

92% believe that auditory alarms were reproduced at a appropriate speed and 92% commented that the auditory alarms sounded were pleasant sounding.

Finally, the questionnaire included a dichotomous question where participants had to decide whether they preferred the auditory alarm with male or female voice. Figure 6 provides the results. The preference for the bias for a female, however, might be due to the much greater percentage of male participants in this study (83%).

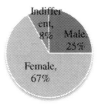

Fig. 6. Preference in the type of voice

5   Conclusions

This paper describes the tests of use of a visual and auditory interface on a driving simulator. Moreover, it reports the results of the usability evaluation of a driving simulator with an auditory and visual interface, providing favorable results regarding acceptability and ease of use. This paper also presented results concerning the usability of different aspects that are considered important in designing visual and auditory interfaces, which was obtained from a questionnaire expressly developed for this purpose. Future work is needed to improve the visual-auditory interface and incorporate it into a real-time system that functions in real-world vehicular ad hoc networks. Since the paper focused on the system design and usability, it did not consider attentional resources to measure their effectiveness in the developed audio-visual warning system.

References

1. OMS. 2nd Global Status Report on Road Safety (2012), http://www.who.int/violence_injury_prevention/global_status_report/flyer_en.pdf
2. Ranney, T.A., Mazzae, E., Garrott, R., Goodman, M.J.: NHTSA Driver Distraction Research: Past, Present, and Future. In: Proceedings of In Driver Distraction Internet Forum (2000)
3. AT&T_Inc. AT&T Natural Voices® Text-to-Speech Demo (2011, Junio), http://www2.research.att.com/~ttsweb/tts/demo.php
4. Noyes, J.M., Hellier, E., Edworthy, J.: Speech Warnings: A Review. Theoretical Issues in Ergonomics Science 7(6), 551–571 (2006)
5. Graham, R.: Use of Auditory Icons as Emergency Warnings: Evaluation Within a Vehicle Collision Avoidance Application. Ergonomics 42(9), 1233–1248 (1999)

6. Sodnik, J., Dicke, C., Tomazic, S., Billinghurst, M.: A user study of auditory versus visual interfaces for use while driving. International Journal of Human-Computer Studies 66, 318–332 (2007)
7. Sodnik, J., Tomazic, S., Dicke, C., Billinghurst, M.: Spatial Auditory Interface for an Embedded Communication Device in a Car. In: 2008 First International Conference on Advances in Computer-Human Interaction, pp. 69–76 (2008)
8. Garzon, S.R.: Intelligent In-Car-Infotainment System: A Prototypical Implementation. In: 2012 8th International Conference on Intelligent Environments (IE), pp. 371–374 (2012)
9. Man Ho, K., Yong Tae, L., Joonwoo, S.: Age-Related Physical and Emotional Characteristics to Safety Warning Sounds: Design Guidelines for Intelligent Vehicles. IEEE Transactions on Systems, Man, and Cybernetics, Part C: Applications and Reviews 40, 592–598 (2010)
10. Cuong, T., Doshi, A., Trivedi, M.M.: Investigating pedal errors and multi-modal effects: Driving testbed development and experimental analysis. In: 2012 15th International IEEE Conference on Intelligent Transportation Systems (ITSC), pp. 1137–1142 (2012)
11. Horrey, W.J., Wickens, C.D., Consalus, K.P.: Modeling Drivers' Visual Attention Allocation while Interacting with In-vehicle technologies. Journal of Experimental Psychology: Applied 12(2), 67 (2006)
12. Horrey, W., Wickens, C.D.: Driving and Side Task Performance: The Effects of Display Clutter, Separation, and Modality. Human Factors: The Journal of the Human Factors and Ergonomics Society 46(4), 611–624 (2004)
13. ISO_9241-11, Norma ISO 9241-11: Ergonomic requirements for office work with visual display terminals (VDTs) - Part 11: Guidance on usability. In: ISO 9241-11, ed. Ginebra, Suiza: Organization for Standardization (ISO) (1998)
14. Brooke, J.: SUS: A quick and dirty usability scale. In: Jordan, P.W., Thomas, B., Weerdmeester, B.A., McClelland, A.L. (eds.) Usability Evaluation in Industry. Taylor and Francis, London (1996)
15. G27. Equipment driving simulator G27 Logitech Racing Wheel (2013),
    http://www.logitech.com/en-us/product/
    g27-racing-wheel?crid=714
16. Racer. Website of the racing simulator of vehicles Racer (2013),
    http://www.racer.nl/
17. LED_Flasher, FLASHER CIRCUITS (2013)
18. Hsien-Tang, L.: Applying location based services and social network services onto tour recording. In: 2012 International Joint Conference on Computer Science and Software Engineering (JCSSE), pp. 197–200 (2012)
19. Bangor, A., Kortum, P., Miller, J.: Determining What Individual SUS Scores Mean: Adding an Adjective Rating Scale. Journal of Usability Studies 4, 114–123 (2009)
20. Bangor, A., Kortum, P., Miller, J.: An Empirical Evaluation of the System Usability Scale. International Journal of Human-Computer Interaction 24, 574–594 (2008)

Using Map Representations to Visualize, Explore and Understand Large Collections of Dynamically Categorized Documents

Ernesto Gutiérrez, J. Alfredo Sánchez,
and Ofelia Cervantes

Universidad de las Américas Puebla, Puebla, Mexico
{ernesto.gutierrezca,alfredo.sanchez,ofelia.cervantes}@udlap.mx

Abstract. This paper presents VOROSOM, a novel visualization scheme that supports collection understanding and exploration of large, distributed collections. Using metadata harvested from diverse collections, VOROSOM produces a map representation in which regions are associated with categories of documents. The shape of each region in the map reflects the relationships among documents in each of the categories. Thus, the distance between two regions directly corresponds to their semantic affinity. Maps are produced in such a way that the number of categories is maintained within a manageable size, considering the user's cognitive capabilities. Maps are organized hierarchically, which supports the exploration and navigation within categories and subcategories of documents using map representations consistently. We report initial results of user studies with a prototypical implementation of our visualization scheme over an actual network of digital libraries.

Keywords: Information visualization, collection understanding, self-organizing maps, Voronoi diagrams, map-based visualization.

1   Introduction

It is now commonplace for experts and the general public to refer to the challenges of ever-increasing volume and complexity of available collections of digital documents. Digital libraries represent enclaves that provide some organization and facilitate access to curated, validated collections and provide tools for supporting knowledge-intensive tasks. However, collections held by digital libraries continue to be vast, complex and, more often than not, components of even larger repositories managed in multiple, distributed servers.

In order to take full advantage of these collections, it is helpful for users not only to be able to search for and retrieve specific items, but also to understand collections as a whole, which implies to be able to obtain an overview of the various facets of collections, as well as the relationships among their attributes. These facets may include, for example, the ontology of subjects on which a collection includes documents, or the authors of documents and their affiliations. Relationships among these attributes may be explicit, as those indicated by lists of co-authors, but others may not be as evident,

C. Collazos, A. Liborio, and C. Rusu (Eds.): CLIHC 2013, LNCS 8278, pp. 140–147, 2013.
© Springer International Publishing Switzerland 2013

such as the overlaps in the lists of keywords or other metadata that describe documents.

Information visualization supports collection understanding by aggregating large collections and their explicit and implicit relationships into meaningful geometrical representations. Many approaches have been designed to address this challenge. We present VOROSOM, a visualization scheme that is based on a map representation in which regions are associated with document categories and are shaped so each region they reflect relationships among documents. Maps are organized hierarchically to also support exploration and navigation using map representations consistently.

2   Related Work

In the literature, diverse approaches address the visualization of large document collections: point spatialization, network, hierarchical, cluster, topic/tag-based , temporal and landscape visualizations. Each of these visualizations attempt to raise awareness of the content, structure, relationships or changes in an information space. Nevertheless, not all visualization interfaces incorporate all four aspects in an integrated fashion. Whereas some interfaces only focus on just one aspect others aim to achieve a complete overview of available information through single or multiple views [1]. In this regard, we emphasize the need for a visualization to provide a comprehensive overview that facilitates the awareness of all four aspects: content (subject matter of the collection), structure (patterns, trends, distribution and size), relationships, and changes (evolution of the collection) of the information space.

Through navigation of information spaces, users gain a better understanding of their contents by traversing their structure and existing relationships. Thus it is desirable for visualization interfaces to provide mechanisms that facilitate navigation and exploration, such as zooming, brushing, and filtering [2].

Self-organized maps (SOMs) are commonly used to visualize high dimensional data, since SOMs synthesize, classify and maintain relationships among elements in an information space. WEBSOM, SOMLib and ViBlioSOM, discussed in [2], have been used to visualize and classify document collections. However, these visualizations are still tied to classic SOM visualizations such as grid, points and U-matrix and typically impose a high cognitive load on the user. Skupin & Fabrikant [3] improved SOM visualization by adding Voronoi segmentation to SOMs grids, but they have troubles with spatialization of points within a single neuron. A different approach is Voromap [4] a visualization tool over IDMAP (a set of combined projection techniques to build document maps). Voromap performs Voronoi segmentation over a set of points created by IDMAP, however there is no information about color coding and labeling of documents. Moreover Voromap approach is limited by the number of documents mapped. In the following section we discuss novel segmentation techniques that can be exploited in order to facilitate the spatialization of SOMs and also can improve the comprehension of the inherent characteristics of SOMs such as structure, relationships and clustering altogether with the aggregation of comprehensive overview and navigation tools. A thorough literature review can be found in [2].

3   The VOROSOM Visualization Scheme

3.1   Conceptual Design

Our main goal is to provide the user with a comprehensive overview by integrating visual elements that help the user understand the content, structure, relationships and changes in document collections. This is accomplished by an interface that incorporates elements to help navigate and explore the collections.

Hierarchical and Dynamic Categorization. In general, human cognitive capabilities cannot cope easily with the rapid growth and complexity of available document collections. An effective manner of assisting people in processing document collections is by simplifying them through classification.

A self-organized map (SOM) is a well-known type of artificial neural network that has the capability of presenting an overview of an information space in a limited visualization area (typically 2-D). Furthermore, SOMs present this overview as consisting of categories, since a SOM is essentially a classification algorithm. In addition, SOMs preserve the topology, i.e., *the relationships and semantic proximity between elements are maintained as they appeared originally in the information space.* Hence, a SOM has three main characteristics useful for visualization and collection understanding: it provides an overview of the collections, it facilitates document categorization, and it preserves the topology implied by existing explicit and implicit relationships among elements in the collections.

Document collections have a hierarchical structure, e.g., in a typical library collection, documents are grouped into subcollections (general topics) and for each subcollection there are even smaller subcollections (subtopics). Following this notion, we integrated an enhanced variation of the SOM algorithm named GH-SOM [5] to perform the dynamic categorization as a hierarchical document structure.

Category Visualization. In order to facilitate the visualization of SOM characteristics, we apply a combined approach of spatial location of SOM categories using 2-D trilateration, as well as their spatial segmentation using Voronoi regions. This process generates Voronoi diagrams over Self-Organized Maps (VOROnoi + SOM, hence VOROSOM) where each Voronoi region is a category, and each category is spatially located to reflect its semantic relationship with neighboring categories. This process is shown in Figure 1, in which we refer to the final Voronoi diagram as "the map".

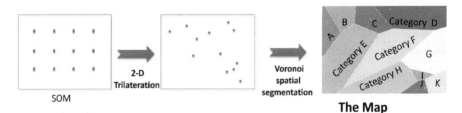

Fig. 1. Visualization of SOM through spatialization and Voronoi regions: The Map

3.2   The Map Metaphor

One of the four major processes to gain insight while using a visualization system is the use of metaphors that match an easy mental model [6]. In this sense, we take advantage of the map metaphor and the hierarchical structure of the dynamically categorized document collections to design an interface that users can rely on to navigate, explore and understand the collection's contents.

Overview. With Voronoi regions as the basis for visualizing SOMs, the map is able to display the content, structure and relationships of document collections and provide a comprehensive overview. The map is thus the key interface component. Categories are color coded and then labeled using Term Frequency Inverse Document Frequency (TFIDF) over the set of documents that each category holds and then selecting the most representative words to describe them. A commonly accepted guideline indicates that, the maximum number of categories that users can easily distinguish based on color coding is twelve [2]. Figure 2 shows the main view of a VOROSOM visualization, in which nine different Voronoi regions represent document categories coded by color and located according to their semantic relationships (as produced by SOM).

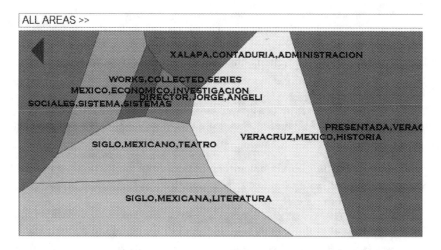

Fig. 2. Main view of VOROSOM visualization

Navigation. The hierarchical structure inherent to document collections and their corresponding classification with GH-SOM also makes hierarchical navigation a natural choice. In this sense we take advantage of the "map metaphor" in order to navigate through the collections. We image our main view as the Pangaea continent and each category of the map, as a country. Then, we can navigate down in the hierarchy to reach smaller regions as states and so on. Figure 3 illustrates some navigation sequences over the hierarchical structure of document collections.

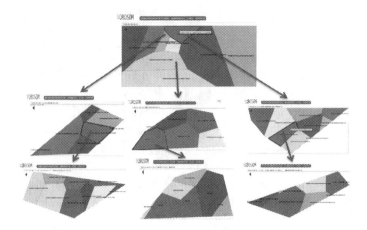

Fig. 3. Navigating the hierarchical structure using the "map metaphor"

4   Implementation

In order to demonstrate the viability and potential of the VOROSOM scheme as a visualization mechanism to facilitate the exploration and understanding of large document collections, a prototype was implemented on top a federation of repositories known as the ReMeRI[1] collections, the main goal of which is to give visibility to scientific work conducted at Mexican universities and the educational resources they have developed.

The VOROSOM[2] prototype consists of three main components: dynamic classification, web interface and document retrieval. Details of VOROSOM implementation can be found in [2].

- *Dynamic classification.* We implemented the dynamic classification using GH-SOM. We retrieve from documents the three most representative attributes (title, subject, description).
- *Web interface.* The visualization is processed and displayed on a web browser using SVG and DOM manipulation using Data-Driven Documents (D3JS[3]).
- *Document retrieval.* It is achieved by querying the database where Dublin Core ® metadata files are located.

In the map of first level (Figure 4a), we can find all areas (or topics in the ReMeRI collections). If we select the region "XALAPA ACCOUNTING ADMINISTRATION", we are actually navigating down to another map (Figure 4b) that contains subtopics of such region. Then, if we select "BUSSINESS ADMINISTRATION" we navigate down to a third level map

---

[1] ReMeRI portal - http://www.remeri.org.mx/ Retrieved May 20, 2013

[2] VOROSOM http://ict.udlap.mx/remeri/VOROSOM1/index.html/
Retrieved June 6, 2013

[3] Data-Driven Documents - http://d3js.org/ Retrieved May 20, 2013

(Figure 4c) that contains subtopics of the previous subregion. Finally, if we point and click on the region "INTERNATIONAL BUSINESS ADMINISTRATION" we can find a set of results (Figure 4d) that represent documents semantically related to our navigation, i.e., we found documents related to international business administration in the Xalapa region.

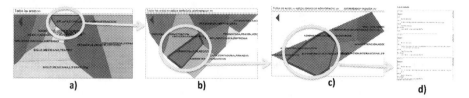

a)   b)   c)   d)

Fig. 4. Navigating with VOROSOM

5   Evaluation

In order to evaluate the VOROSOM prototype interface, a qualitative study was designed and users were observed as they performed specific tasks. A total of eight users (two females and six males), ages between 21 and 25, participated in the study. All subjects were undergraduate students majoring in areas related to Information Technologies. Participants received extra grade points for a Human-Computer Interaction course as compensation.

Sessions were conducted in a usability laboratory using the constructive interaction methodology, in which pairs of users interact with each other and with the interface. Thus four couples were arranged (F-M, F-M, M-M, and M-M). A facilitator guided the session while sitting next to the participants.

5.1   Procedure

Prior to the study, a preliminary questionnaire was applied in order to obtain details from participants for further analysis of study results. Once questionnaires were applied, participants performed tasks in four different groups:

- Group 1: *Observation of graphic elements of the interface.* This group focused on detecting what users understand, without explanation, about visual elements.
- Group 2: *Navigating the collections.* This group aims to detect the degree of difficulty entailed by navigating over the "map metaphor" represents.
- Group 3: *Exploration of collections.* In order to explore collections users need a certain degree of understanding, in this way, this group of tasks helps to determine if the interface facilitates collection understanding.
- Group 4: *Free exploration of collections.* When users already have an overview of collections they can explore their contents with other objectives such as discovering and learning. This group of tasks helps us determine whether users reach these goals.

At the end of the study, another questionnaire was applied in order to find issues that might not have been verbalized by the participants during the usability test.

5.2   Results

Preliminary questionnaires revealed that users had no prior knowledge on Voronoi diagrams neither on SOMs. On the other hand, they did have previous knowledge on seeking documents in large collections and were enthusiastic about research tasks.

Post-test questionnaire results are summarized in Figure 6, where a total of eight aspects were evaluated. The scale of user responses for each aspect was {1:Totally Disagree, 2:Disagree, 3:Agree, 4:Totally Agree}, so a maximum of 32 can be reached if responses from eight users are added up. Results indicate that VOROSOM is useful for collection understanding as it presents an overview and facilitates retrieval of information semantically related. This is achieved through (hierarchical) navigation of collections in an easy and intuitive way (using a map metaphor). Also, the visualization interface allows the user to discover information.

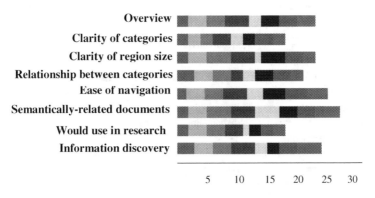

Fig. 5. Results of post-test questionnaires

Users verbalized and wrote down comments about their impressions over the interface. Related to semantic relationships among documents, one user expressed: *"interface allows you to find documents related to each other thanks to the regions (representation)"* and another user complemented *"regions allow me to explore, in a better way, topics and subtopics"*. Related to collection understanding, one user commented: *"It facilitates the exploration of digital collections to people who are not familiar with them"* whereas another said *"it can be used for learning with children"*. Lastly, with respect to the ease of use one user said: *"All you have to do is check that subjects want and to click to get to the documents of interest"*.

Even though VOROSOM was well evaluated, users had trouble to understand labeling due to the combination of nouns, verbs and adjectives. Moreover, it is still needed to implement filters and details on demand. Despite those issues, users completed most tasks successfully.

6   Conclusions

We have introduced a novel visualization scheme aimed to support collection understanding. Our scheme, referred to as VOROSOM, relies on a map metaphor to represent document categories that are inferred dynamically as well as the relationships among the underlying documents. The map representation is used consistently to allow users both to obtain an overview of the collections and to navigate and explore each of the categories. Our initial evaluation shows that users are able to grasp easily the major subjects and the relationships among documents represented using maps based on Voronoi diagrams that are constructed on top of self-organized maps.

Our user studies also have allowed us to find various areas for improving and advancing the implementation of the VOROSOM visualization scheme. We plan to implement and evaluate alternative map representations including the use of curved rather than straight lines as well as non-contiguous regions that depict categories more clearly. Moreover, we are developing a new way to label map regions so they will become tag clouds in which keywords are presented on a larger font if they refer to a larger number of documents in the corresponding categories. Finally, we will continue to work on improving filtering components at the interface level, as well as to provide functionality to support analysis of collections over time.

Acknowledgements. This project has received support from CONACYT through the ReMeRI project (www.remeri.org.mx). We also acknowledge support from the French-Mexican Laboratory of Computer Science and Automatic Control (LAFMIA-CNRS UMI 3175).

References

1. Zhang, Y., Li, T.: DClusterE: A Framework for Evaluating and Understanding Docu-ment Clustering Using Visualization. ACM Trans. Intell. Syst. Technol. 3(2), 24:1–24:24 (2012)
2. Gutiérrez Corona, E.: Visualización de información para entendimiento de colecciones usando VOROSOM, M. Sc., Universidad de las Americas Puebla (2013)
3. Skupin, A., Fabrikant, S.I.: Spatialization Methods: A Cartographic Research Agenda for Non-geographic Information Visualization. Cartogr. Geogr. Inf. Sci. 30(2), 99–119 (2003)
4. Pinho, R., de Oliveira, M.C.F., Minghim, R., Andrade, M.G.: Voromap: A Voronoi-based tool for visual exploration of multi-dimensional data. In: Proceedings of the Conference on Information Visualization, Washington, DC, USA, pp. 39–44 (2006)
5. Rauber, A., Merkl, D., Dittenbach, M.: The growing hierarchical self-organizing map: Exploratory analysis of high-dimensional data. Trans. Neur. Netw. 13(6), 1331–1341 (2002)
6. Yi, J.S., Kang, Y., Stasko, J.T., Jacko, J.A.: Understanding and characterizing in-sights: how do people gain insights using information visualization? In: Proceedings of the 2008 Workshop on Beyond Time and Errors: Novel Evaluation Methods for Information Visualization, New York, NY, USA, pp. 4:1–4:6 (2008)

Augmenting Decision Tree Models Using Self-Organizing Maps

Wilson Castillo-Rojas[1], Fernando Medina-Quispe[1],
and Claudio Meneses-Villegas[2]

[1] Faculty of Engineering and Architecture, Arturo Prat University, Iquique, Chile
{wilson.castillo,femedina}@unap.cl
[2] Department of Systems Engineering and Computing, Northern Catholic University,
Antofagasta, Chile
cmeneses@ucn.cl

Abstract. This study considers the application of the Self-Organizing Map technique on a decision tree model generated to achieve model-augmented visualization, based on a visual perception model scheme called VAM-DM. It supports the visual analysis of a data mining model in the adjustment phase, also combining complementary views of graphical artifacts for each component or node of the decision tree. It seeks to answer user generic questions regarding the model inner workings and to achieve a better understanding of the model finally obtained. In this context, the Self-Organizing Map technique serves a dual purpose: spatial partition of the data subset associated with a tree node and partition visualization with a map. Finally, a controlled experiment is carried out with a software prototype and two user groups, novices and experts in DM's processes, and results from this experiment are analyzed. This analysis allows us to assess the usefulness of the Self-Organizing Map technique for augmented decision tree model and their efficiency to support the comprehension of the generated model.

Keywords: Data Mining, Visualization, and Visual Exploration of Data Mining Models.

1 Introduction

The process of Knowledge Discovery in Databases (KDD) is complex, and many obstacles, research questions and problems need to be investigated and clarified. An important aspect is the understandability of the entities involved in the process Data Mining (DM) itself. When users and data analysts wish to interact with these entities to improve outcomes, they need more than just input/output information, what they really require is to understand how entities are internally working, its components, and the process carried out by them and how they relate to each other. In this context, the visualization paradigm has been applied in a very limited way in the KDD process and mainly focused on data visualization (process input) and result visualization (process output). Regarding this, appropriate visualizations applied to DM models can transform these into understandable tools that convert data into knowledge.

C. Collazos, A. Liborio, and C. Rusu (Eds.): CLIHC 2013, LNCS 8278, pp. 148–155, 2013.

To achieve this, our proposal is based on a scheme of Augmented Visualization for Data Mining Models called VAM-DM. This work includes the implementation of a VAM-DM scheme on prototype software that lets users generate models from a selected primary DM technique, using a set of appropriate previously prepared data. Afterwards, the user may select a secondary DM technique for visual augmenting, and can apply it to different components or nodes of the decision tree generated, including its root node. Also, users dispose of a set of visual elements that can be applied to the data of the selected nodes, among which are some traditional graphical artifacts. In addition, this tool provides different mechanisms of interaction that allow users to navigate and explore the model and its components in a single interface without losing the context, achieving augmented visualization for DM models.

In this paper, a decision tree technique is combined with a self-organizing map in order to illustrate the main idea of this work. The use of visualization of a data mining model can improve the model understanding and therefore the usefulness of the KDD whole process. Decision tree learning is one of the most widely used and practical methods for inductive inference. It is a method for approximating discrete-valued functions that is robust to noisy data and capable to learn disjunctive expressions. Among the most known decision tree algorithms are CART, ID3 (C4.5, J48, C5.0), CHAID, for example. On the other hand, self-organizing maps are competitive learning algorithms that organize the output information in a map, allowing an easy visualization of the spatially partitioned input dataset.

Finally, a subjective evaluation based on a software prototype is presented through the development of a controlled experiment, including a survey of two groups (novice and expert) of users/analysts. They used a previously prepared dataset together with the software prototype and the WEKA tool, in order to perform a predefined DM task designed for this purpose, and provided information about their performance, usability, handling visualization and support in understanding the DM model. The results and their analysis allow us to validate the proposal and scientific contributions of this part of our research.

2 Visualization in Data Mining

Visualization is increasingly being incorporated in the KDD process as a tool to support the interactions between users, data analysts and the components involved in the development of DM process. However, none of the existing DM process models [1] [3] [5] incorporate and discuss the role of visualization in the DM cycle. Meneses [7] propose a scheme of DM process with support to visualize four types of entities: data, parameter space of DM algorithms, induced models and patterns. Data visualization supports the interaction between users or data analysts and datasets involved in the DM process. For example, visualization can be used to obtain a preliminary understanding of the data and refine the objectives and tasks defined initially by the user in the problem formulation phase. Some of these techniques are limited to deal with sets of low-dimensional data (e.g., scatter-plots), while others focus on datasets of high dimensionality (e.g., parallel coordinates, iconographic representations, visualizations radial, Chernoff faces). Keim [6] provides a comprehensive taxonomy of visual techniques to explore massive datasets, while Hoffman [4] outlines a categorization of techniques for visual DM tasks. On the other hand, model visualization supports understanding and interaction with the model induced from a set of training data by a DM algorithm [2] [8]. These visualizations should provide a natural way to understand the structure, components, and complexity of a model, as well as visual representations provide a direct way to compare several models, and allow the use of human visual perception to formulate hypotheses and conclusions about the model and its correlation with the data. Model visualizations have recently appeared in the relevant literature and incorporated into some software tools of data analysis for DM. Patterns visualization is referred to visualize results of applying a DM induced model to a validation and/or testing data set [8] [9]. For predictive models, these results are commonly given as error rates or a confusion matrix. In this case, visualization can be used to support the interpretation of these results, and to provide visual feedback to correlate these patterns with data, parameters, and models used to generate them.

3   Visual Perception and User Interaction Reference Models for Data Mining Tasks

First, a reference model Visual Data Exploration proposed by Keim et al. [11] is considered, which suggests that the process of Visual Analysis (VA) is characterized by interactions between data, views, models about the data, and users in order to obtain knowledge. In this model, the KDD process is established in different states, from the preparation phase of the data, through exploration, modeling, visualization to support comprehension and validation of the model, to the stage of obtaining knowledge. Also shown are the various relationships and actions between states, all within the framework of an iterative process. The second model analyzed corresponds to an Interactive Display Concept Model of Data Mining proposed by Yan Liu [12]. This model is more specific than the Keims reference model, and it details the actions taken by both the machine and the user, in both cases associated with a logic of interaction related to the construction and evaluation of the DM model, i.e. it does not take the previous phases (preparation and processing of data) and later ones (knowledge acquisition). A third model analyzed is the one proposed by Vitiello [10], which integrates Visual Analytics based on a workflow for developing senses (sensemaking), evolved from naturalistic decision making research from Klein [16]. The key issues in this scheme are related to pattern recognition, also referred as frames or boxes. These can be seen as a mental map of the situation informing the decision. Tables can also be viewed as analogous to the fluid hypothesis, in that these can be developed when more information becomes available.

4   The VAM-DM Proposed Approach

The VAM-DM scheme proposed in [15] brings the concept of "Augmented Visualization" for DM models, and what it proposes is that given a DM technique to visualize called primary data mining technique (PT-DM) in this scheme, it allows users to incorporate to this view different ad-hoc visual elements to the model and the data domain, and in turn to apply another DM technique called secondary data mining technique (ST-DM) as visual augmenter that allows users to explore the DM model induced from the PT-DM. The ST-DM selected technique must meet three requirements: being a descriptive technique, appropriate to the data which is working the PT-DM, and to provide additional information of the model generated by the PT-DM. Additionally, the proposed scheme provides an appropriate set of mechanisms for user interaction. This set of visualizations points to achieve higher Visual Analytics of the model in its stage of refinement or adjustment. This combination of DM techniques is proposed by several authors as a mechanism to better understand not only the data but also on the models generated. The architecture of the VAM-DM scheme is represented in general terms in Figure 1.

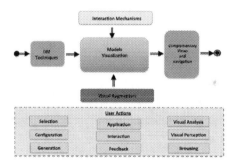

Fig. 1. VAM-DM Scheme [15]

This figure also shows the different components of the VAM-DM scheme, starting from a set of DM techniques, which may be primary or secondary techniques. Also, a set of visual augmenting tools is established, among which include traditional graphical artifacts, as well as DM techniques selected to act as a visual browser of other technique. In addition, the VAM-DM scheme includes different interactions that users can perform in this process.

5   Case Study

One of the key problems of DM techniques is its visual representation and understanding the inner workings of the model for the user/analyst, which in the case of DT is more complex for: large trees, dataset to be analyzed has a high dimensionality, and due their hierarchical structure characteristic [13]. From the comparative review of visualization schemes for DM techniques [14] is concluded that: 1) most of the research works recommend using an appropriate combination of visualization schemas with different DM techniques, depending on the task to be performed and the data characteristics; 2) it is essential to consider the incorporation of user interaction mechanisms in the design of new visualizations; 3) the visualization role in the KDD process must be understood and extended in all its stages, so that allows exploring the data, models and patterns obtained.

Particularly, for the DT technique, most visual representations discussed in [14] proposed a normal form hierarchical dendrogram at a static display, without the possibility that the user/analyst can interact with each one of the nodes of the tree. Most revised DM visual tools, although they deliver general tree information along with the associated confusion matrix, do not allow combining DM techniques to provide information beyond the rules of the model and instances at each node, such as information about data dispersal and spatial distribution at each node, much less provide user interaction mechanisms, in order to browse, select and explore each tree component or node. Using the Self-Organizing Map (SOM) technique applied to a DT as a visual enhancer has a dual purpose: spatial partition of the subset of data associated with each tree node, and display of this partition using a map. Whereas the tree by itself allows setting decision rules by distributing data across hierarchies represented by their nodes, and each node collects the instances that comply with these rules, however, it does not allow visualizing the spatial distribution of the instances, which does provide SOM. In addition, the SOM technique is appropriate to the domain of the data handled by the DT, and to describe their distribution in each node. This allows a comparison between nodes and thus determining those with similar distribution or large differentiation, through the specification of the distance, the number of instances located over or under the centroid of a grid map.

The visual environment prototype developed for experimental analysis considers the implementation of VAM-MD scheme for the hierarchical decision tree model as the PT-DM technique, in combination with the SOM technique as visual augmenter for exploration and analysis. It also incorporates a set of visual or graphical artifacts and different mechanisms of interaction (zooming applied to visual elements and visual augmenter, handling of transparency to maintain tree context, selection of nodes at each level, and tree explorer).

Figure 2 shows an overview of the experimental prototype main interface, which displays a DT in the center, with additional views and visual elements on the right. In this interface, all components of the visual tool can be observed: a) Selecting and setting parameters of the PT-DM to visualize and generation of a DM model, in this section the user can select the dataset to be analyzed, and then display some characteristics (e.g., number of instances, type and number of attributes). Also, the user can select the primary technique of data mining. Then, once configured with the PT-MD parameters selected, users can run this algorithm or technique in order to generate the DM model. b) Visualization area of the primary DM technique. This is the main work area which shows an

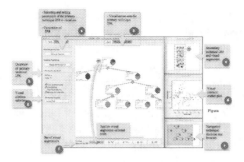

Fig. 2. Principal interface of prototype software

overview of the selected PT-DM, and where users can interact (browse, explore and select) with each element or node of the tree. c) Secondary DM technique as a visual augmenter. In this area, the minimized view of the ST-DM is showed and user can select and apply it to the PT-DM. Also in this area, the user can maximize the view, allowing changing its settings. For this work, the SOM technique has been considered as visual augmenter ST-DM. d) Visual element. In this area are deployed visual elements selected beforehand. In this area, a scatter plot of the selected node can be observed. e) Navigation technique. It corresponds to a browser of the PT-DM, and allows the user to have the guidance of the area or level where is crossing the DT. f) Visual augmenter bar. It lies at the bottom of the tool working area and provides a set of visual augmenters or ST-DM techniques. g) Selection of visual elements. In this section the user has an appropriate set of graphical artifacts which can select, configure and apply to the PT-DM. Within the set of visual available elements are: histogram, mosaic and scatter-plots. h) Overview of the PT-MD. Here, generic information for PT-DM, number of nodes and leaves in the DT are shown. For each selected node, including the root node, the user can apply SOM to visualize the distribution of the items on the tree nodes, which also provides user interaction elements to change the type of map background color and the type of class, the chart type of each item on the map, to select the type of class to color in the representation, and select the test or training set for visualization.

6   Evaluation and Results Analysis

In order to assess the usefulness of the proposed approach, a controlled experiment is carried out. This controlled experiment provides benchmarking between the software prototype and WEKA software in performing a DM task predefined for this purpose, to generate a DT model with a dataset previously prepared. This experiment was performed with a universe of 13 people with different levels of knowledge about data mining, the use of data mining tools, and in particular about the WEKA software. These people were categorized into two groups of users: novice and expert. Using both tools, users participating in this experiment had to generate a DT model, visualize, interact with the model, and interpret patterns or rules obtained using the two software options available. The goal established for participants was to make a generic classification job and answer questions about the model, its components, and relate the model to the characteristics of the data from which the model was generated by the DT algorithm used. Subsequently, once the DM task prepared for this experiment was performed, users/analysts provided their subjective judgments, captured through a survey designed for this purpose, regarding: the performance of both tools at hand, management of the visualization of the DT generated model, usability, usefulness of visual elements provided, the appropriateness of combining SOM and apply to a DT model achieving and visually augmented model, and efficiency in understanding the generated model.

Overall, for both groups (novice and experts), the prototype software was widely accepted, from the point of view of usability and performance of this tool, declaring in a 100 % that this tool allows to find some kind of relationship between the attributes of the dataset. Compared to the WEKA software, prevails a positive assessment of the degree of usefulness of the options and parameters offered by the software prototype, in order to understand the DT model generated. In the case novices about 83 % and 17 % of them considered the software prototype usefulness high and very high respectively, while 80 % and 20 % of the experienced users considered it high and regular, respectively. The utility measures of the set of views provided by the experimental prototype to understand the partitions made by the DT algorithm was considered high for 69% of users in general, and very high for 15 % for expert users. However, this last one group the 15 % considered low or regular this utility measure. The ability to describe the data in a node by using the technique SOM, to deliver enhanced visualization of the DT, given to both groups of users is very high and high with 39 % and 61 %, respectively. With this we can assess that the combination and application of the SOM technique on a DT allows, by one hand, to complement the model visualization generated from the DT, and, on the other hand, to improve the understanding of the model. The degree of acceptance of the software prototype, according to users experience with other data mining tools, is high considering these two groups.

7   Conclusions and Future Work

The preliminary findings obtained in this work are the following. It was possible to confirm the appropriateness and usefulness of combining the SOM technique on a DT model previously generated, as a complementary technique to visualize and describe the instances on each node of the tree, bringing spatial data visualization through a map. It was observed as experimental result and the data obtained in the survey, a trend of both expert and novice users a high valuation of using SOM technique complementary to describe the DT components. Also, users observed that the application of SOM on a DT can improve the understanding of DM models over other tools. In addition, the provision of visual or graphical artifacts incorporated in the software prototype, applied to the data in each DT node, meet to support the analysis and exploration of the generated model. With regard to future work, authors are currently working on the following aspects: other descriptive DM techniques are being evaluated, to provide additional views besides the SOM technique to augment visually a model of DT. Also, the inclusion of additional graphical artifacts that may be more useful in exploring the data for each node of a tree is being considered, preferably in conjunction with high-dimensional data. Finally, the extension of the prototype software with additional interaction mechanisms for comparing nodes of a DT through the map provided by the SOM technique, such that quantitatively measure the degree of similarity between these nodes through the maps.

References

[1] Adriaans, P., et al.: Data Mining. Addison-Wesley (1996) ISBN 0201403803
[2] Becker, B., Kohavi, R., Sommerfield, D.: Visualizing the Simple Bayesian Classifier. In: Workshop on Issues in the Integration of Data Mining and Data Visualization. Springer (1998)
[3] Fayyad, U.M., Piatestky, S.G., Smyth, P.: The KDD Process for Extracting Useful Knowledge from Volumes of Data. Comm. ACM 39(11), 27–34 (1996)
[4] Hoffman, P.E.: Table Visualizations: A Formal Model and its Applications. Sc.D. Thesis, Dept. of Comp. Science, University of Massachusetts at Lowell (1999)
[5] John, G.H.: Enhancements to the DM Process. Stanford University (1997)
[6] Keim, D.: Visual Techniques for Exploring Databases. In: Tutorial Notes in the Third International Conference on KDD, Newport Beach, CA (1997)
[7] Meneses, C., Grinstein, G.: Visualization for Enhancing the DM Process. In: Proceedings of the DM and KDD: Theory, Tools, and Technology (2001)
[8] Thearling, K., Becker, B., DeCoste, D.: Visualizing Data Mining Models. In: Proceedings of the Integration of DM and Data Visualization Workshop (1998)
[9] Humphrey, M., Cunningham, S.J., Witten, I.H.: Knowledge Visualization Techniques for Machine Learning. Intelligent Data Analysis (2), 333–347 (1998)
[10] Vitiello, P.F., Kalawsky Roy, S.: Visual Analytics: A Sense-making Framework for Systems Thinking in Systems Engineering (2012) ISBN: 978-1-4673-0748-2
[11] Keim, D., Kohlhammer, J., Ellis, G.: Mastering the Information Age Solving Problems with Visual Analytics. Eurographics Association Postfach 8043-38621 (2010)
[12] Liu, Y., Salvendy, G.: Visualization support to better comprehends and improves DT classification modelling process (2007) ISSN 1463922X
[13] Castillo, W., Meneses, C.: Graphical Representation and Exploratory Visualization for Decision Trees in the KDD Process (2012) ISBN 978-1-4673-0793-2
[14] Castillo, W., Meneses, C.: A Comparative Review of Schemes of Multidimensional Visualization for DM Techniques. III INFONOR-CHILE, Chile (2012)
[15] Castillo, W., Meneses, C.: Augmented DM Models Using Visualization (2013)
[16] Klein, G.: A Recognition-Primed Decision Model of Rapid Decision Making. Decision Making in Action: Models and Methods 5(4), 138–147 (1993)

Author Index